Charlotte Ockeloen

Split hand/split foot malformation

Charlotte Ockeloen

Split hand/split foot malformation

Determining the frequency of genomic aberrations with molecular-genetic methods

Südwestdeutscher Verlag für Hochschulschriften

Impressum/Imprint (nur für Deutschland/ only for Germany)
Bibliografische Information der Deutschen Nationalbibliothek: Die Deutsche Nationalbibliothek
verzeichnet diese Publikation in der Deutschen Nationalbibliografie; detaillierte bibliografische
Daten sind im Internet über http://dnb.d-nb.de abrufbar.

Alle in diesem Buch genannten Marken und Produktnamen unterliegen warenzeichen-, marken-
oder patentrechtlichem Schutz bzw. sind Warenzeichen oder eingetragene Warenzeichen der
jeweiligen Inhaber. Die Wiedergabe von Marken, Produktnamen, Gebrauchsnamen,
Handelsnamen, Warenbezeichnungen u.s.w. in diesem Werk berechtigt auch ohne besondere
Kennzeichnung nicht zu der Annahme, dass solche Namen im Sinne der Warenzeichen- und
Markenschutzgesetzgebung als frei zu betrachten wären und daher von jedermann benutzt
werden dürften.

Verlag: Südwestdeutscher Verlag für Hochschulschriften GmbH & Co. KG
Dudweiler Landstr. 99, 66123 Saarbrücken, Deutschland
Telefon +49 681 37 20 271-1, Telefax +49 681 37 20 271-0
Email: info@svh-verlag.de
Zugl.: Berlin, Charité Universitätsmedizin, Dissertation, 2010

Herstellung in Deutschland:
Schaltungsdienst Lange o.H.G., Berlin
Books on Demand GmbH, Norderstedt
Reha GmbH, Saarbrücken
Amazon Distribution GmbH, Leipzig
ISBN: 978-3-8381-2554-1

Imprint (only for USA, GB)
Bibliographic information published by the Deutsche Nationalbibliothek: The Deutsche
Nationalbibliothek lists this publication in the Deutsche Nationalbibliografie; detailed
bibliographic data are available in the Internet at http://dnb.d-nb.de.

Any brand names and product names mentioned in this book are subject to trademark, brand
or patent protection and are trademarks or registered trademarks of their respective holders.
The use of brand names, product names, common names, trade names, product descriptions
etc. even without a particular marking in this works is in no way to be construed to mean that
such names may be regarded as unrestricted in respect of trademark and brand protection
legislation and could thus be used by anyone.

Publisher: Südwestdeutscher Verlag für Hochschulschriften GmbH & Co. KG
Dudweiler Landstr. 99, 66123 Saarbrücken, Germany
Phone +49 681 37 20 271-1, Fax +49 681 37 20 271-0
Email: info@svh-verlag.de

Printed in the U.S.A.
Printed in the U.K. by (see last page)
ISBN: 978-3-8381-2554-1

Copyright © 2011 by the author and Südwestdeutscher Verlag für Hochschulschriften GmbH &
Co. KG and licensors
All rights reserved. Saarbrücken 2011

TABLE OF CONTENTS

PREFACE 1

1. INTRODUCTION 3
    1.1   Pathogenesis of limb development 4
    1.2   SHFM1 (MIM 183600) 6
    1.3   SHFM2 (MIM 313350) 7
    1.4   SHFM3 (MIM 600095) 7
    1.5   SHFM4 (MIM 605289) - Mutations in the TP63-gene 9
    1.6   SHFM 5 (MIM 606708) 10
    1.7   Evidence for two new SHFM loci 11
    1.8   Split hand/foot malformation and long bone deficiency 12
    1.9   Genotype/phenotype correlation 13

2. HYPOTHESES 14

3. MATERIALS AND METHODS 15
    3.1 Patients 15
    3.2 Multiplex Ligation-dependent Probe Amplification (MLPA) 18
    3.3 Real-Time quantitative PCR SHFM 3 locus (10q24) 22
    3.4 Array CGH 25

4. RESULTS 29
    4.1 Array CGH results 39
    4.2 Familial cases 41

5. DISCUSSION 44
    Future prospects 48

ZUSAMMENFASSUNG 49

REFERENCES 51

PREFACE

First of all, I would like to thank Dr. rer. nat. Eva Klopocki, for her excellent supervision and guidance throughout the project. She helped me accomplish my goals and was always stimulating and helpful. Second, I would like to thank my laboratory colleagues, Randy Koll and Fabienne Trotier, for their assistance with my experiments and being such nice colleagues.

I owe much gratitude to Prof. Dr. Mundlos, who gave me the opportunity to work on this project. Also, I would like to thank all other colleagues from the genetics department of the Charité who helped or assisted me during my time in the laboratory.

1. Introduction

Split hand/split foot malformation (SHFM), also known as ectrodactyly or cleft hand/foot, is a complex congenital limb defect that is characterized by a deep median cleft with absence of central ray(s). SHFM presents as a non-syndromic entity or as part of a syndrome. It occurs either sporadically or in families. Reduced penetrance is frequently observed, and has been documented in several pedigrees.[4] One of the well-recognized hallmarks of SHFM is the inter- and intra-familial phenotypic variability; limb defects range from minor syndactyly of the digits to severe syndactylous hypoplasia of several digits or monodactyly.[3]

Oligodactyly, presenting as three or more digits in association with syndactyly and a deep median cleft, is by far the most common pattern.[4] The other two core phenotypes are monodactyly and bidactyly, formerly known as "lobster claw" malformation. Noncore phenotypic manifestations include polydactyly, triphalangeal thumb, clinodactyly, camptodactyly, transverse phalanges, and ulnar deviation.[4,6] Approximately 40% of individuals presenting with SHFM have associated non-limb congenital anomalies, for example mental retardation, cleft palate or ectodermal dysplasia. The overall prevalence of SHFM is reported to range from approximately 0.6/10 000 newborns to 0.51/10 000 newborns.[7]

SHFM can be categorized as typical or atypical. This differentiation was originally made by Lange in 1937[8] and has been maintained by others[9,10]. Atypical split hand is usually unilateral, without associated foot involvement, and occurs sporadically. Regarding the nomenclature and classification, there is significant confusion. It has been postulated that atypical cleft hand may be caused by vascular disruption.[13] According to the Committee of the International Federation of Societies for Surgery of the Hand, the term atypical split hand should be replaced by "symbrachydactyly".[11] However, many clinical geneticists continue to refer to this entity as atypical split hand. In typical split hand, bilateral involvement can occur as well as involvement of the feet. Patients may have a positive family history. The split hand/split foot malformation is usually inherited in an autosomal dominant manner, although autosomal recessive inheritance has also been described.[12]

So far 5 different genetic loci have been mapped for non-syndromic SHFM, and recently evidence for two new loci has been found.[17,19,21,22,28,27,39]

1.1 PATHOGENESIS OF LIMB DEVELOPMENT

The formation of the upper limb occurs in the 4th week of embryonic development and is completed approximately 8 weeks later. The initiation of the lower limb bud formation is delayed by 2 days, but the factors that control limb development are the same for both upper and lower limbs. Thus, it is not uncommon that limb abnormalities occur symmetrically.[3]

The outgrowth and patterning of the limb occur in three dimensions: proximo-distal (shoulder-finger direction), antero-posterior (thumb-little finger direction) and dorso-ventral (back-palm direction). The apical ectodermal ridge (AER) is a thickened ridge of ectoderm at the apex of the limb bud; it controls the outgrowth of the limb bud along the proximo-distal axis. Directly underlying the AER is the progress zone (PZ), an area of rapid cell division. Signals from the AER allow the underlying cells of the PZ to maintain their proliferative activity. The zone of polarizing activity (ZPA), which is located in the posterior region of the developing limb bud, controls the antero-posterior patterning of the limb. Dorso-ventral patterning of the limb is controlled by the genes *Wnt7a* and *Lmx1*. Surgical removal of the AER results in truncation of all skeletal elements of the limb (stylopod, zeugopod, and autopod).[3,24,25]

A number of key players in the AER have recently been identified; these include fibroblast growth factors (Fgfs), bone morphogenetic proteins (Bmps), Wnt signalling molecules, and homeobox containing proteins, such as Msx1 and Msx2 (Fig.1). AER formation is induced by mesodermal signalling to the overlying ectoderm, using Fgf10 and Bmps. Bmps control the ectodermal expression of Msx transcription factor genes. The two major functions of Fgfs induce the proliferation of mesenchymal cells in the PZ, and they are required by the ZPA to maintain Sonic Hedgehog (Shh) expression. Sonic Hedgehog mediated Bmp signalling is essential to maintain the AER. This shows that a co-dependence exists between the AER and the ZPA.

Fgfs are crucial for limb development; *Fgf4* and *Fgf8* knockout mice develop a normal AER, but mesenchymal gene expression is disturbed. This results in aplasia of the proximal and distal limb elements.[24]

Several homeobox genes, such as *HoxD* and *HoxA*, are responsible for maintaining the relationship between the AER and the PZ. In addition, Bmp signalling plays an important role in this process. The homeobox genes are also essential for the formation of the individual digits of the fetal hand.[3,24,25]

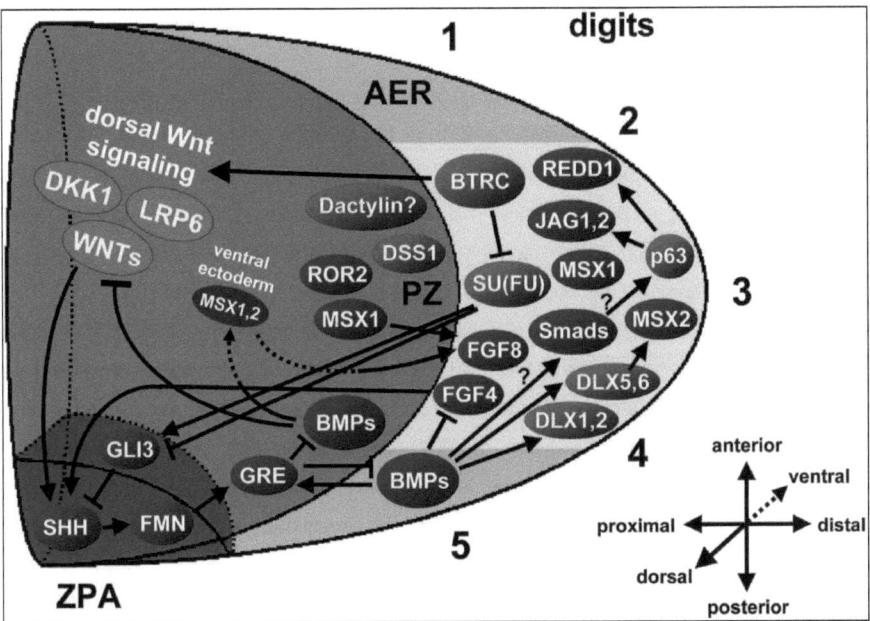

Figure 1. Signalling pathways in the developing limb bud. Failure to maintain the AER or defective AER signalling underlies SHFM. Correct signalling in the anterior and posterior apical ectodermal ridge (AER; light grey), but not in the median AER (yellow), may explain the relatively normal development of the anterior and posterior digits, respectively, while the median digits either develop very poorly or do not form at all. The positions of the AER, the underlying progress zone (PZ; dark grey), and the zone of polarizing activity (ZPA; brown) are indicated. Numbers 1–5 refer to the future positions of digits 1–5, respectively. Directions of the three-dimensional axes are indicated. Protein products from positional candidate genes for isolated SHFM are highlighted in red. Other molecules are shown in blue. Dorsally and ventrally expressed proteins are depicted in lighter and darker blue, respectively. Inhibitory and stimulatory effects are indicated with bars and arrows, respectively. (Adapted from: Duijf PHG et al. Pathogenesis of split-hand/split-foot malformation. Hum Mol Genet 2003; 12: R51-60.)

1.2 SHFM1 (MIM 183600)

Studies with SHFM patients carrying cytogenetically visible chromosome rearrangements have led to the mapping of an autosomal form of the disease to chromosome 7q21-7q22. This locus has been designated SHFM1.[15] By molecular and cytogenetic analysis, a minimal critical region of approximately 1.5 Mb was established. In this region several genes were located which could play a role in SHFM pathogenesis. Two of them are mammalian homologues to the Drosophila distal-less (dll) family: *DLX5* and *DLX6*.[16]
Crackower et al.[17] analyzed six patients with an interstitial deletion at the SHFM1 locus and seven patients with translocations and screened for candidate genes in a 500 kb region containing five of the translocation breakpoints. They identified another candidate gene, designated *DSS1* (also known as *SHFM1*). In the developing mouse limb bud, *Dss1* seems to be expressed predominantly in the limb and facial primordia during early embryonic development. It is also expressed strongly in the dermis of newborn mice, early genital bud and possibly the tooth primordium. When during embryogenesis the expression of this gene is reduced, this could explain not only the SHFM phenotype, but also some forms of syndromic ectrodactyly including Ectrodactyly-Ectodermal Dysplasia-Cleft Lip/Palate (EEC) syndrome (MIM 219900).
Studies of *Dlx5* and *Dlx6* in mice show that these genes are expressed in almost every developing skeletal element and in the forebrain. Their expression patterns are almost identical. In the rat, *Dlx5* showed additional expression in the AER of limb buds.[17]

Deafness is associated with SHFM1 in 35% of the patients and ectrodactyly/deafness has been identified as a distinct clinical disorder (SHFM1D; MIM 220600). SHFM1 is the only locus that involves sensorineural hearing loss, but conductive hearing loss has been associated with EEC syndrome. Tackels-Horne et al.[37] investigated two families with ectrodactyly and sensorineural hearing loss, and mapped these families to the SHFM1 locus at 7q21. Recently, Bernardini et al.[38] reported on a 5-year-old patient with psychomotor delay, ectrodactyly of right hand and both feet, craniofacial dysmorphic features, cleft palate, deafness, and Tetralogy of Fallot. They found a reciprocal interstitial translocation t(7;8)(q21q22;q23q24) with a paracentric inversion of 7q and a microdeletion of 7q21.13, which included the *Fzd1* gene. The deletion found in this patient confirms that the SHFM1D locus maps to 7q21 and suggests new candidate genes for the disorder.

Robledo et al.[36] performed a study with *Dlx5/Dlx6* knockout mice (*Dlx5/6⁻/⁻* mice). The targeted disruption of *Dlx5/6* resulted in bone, inner ear, and several craniofacial defects. It was shown that *Dlx5* and *Dlx6* appear to act as essential regulators of endochondral ossification. Furthermore, *Dlx5/6* control proximo-distal patterning in the murine hindlimb by maintaining the medial portion of the AER, as their loss leads to AER degeneration resulting in a phenocopy of the SHFM1 phenotype with craniofacial defects. In conclusion, *Dlx5* and *Dlx6* are critical regulators of mammalian limb development.

However, so far no mutations could be detected in *DLX5*, *DLX6*, or *DSS1*. Also, none of these genes seem to be interrupted directly by any of the deletion, inversion or translocation breakpoints.[54] The role of these genes still remains unclear, although several hypotheses have been proposed to explain how deletions and translocations at the SHFM1 locus could cause the SHFM phenotype. One theory is that distant *cis*-acting regulatory elements are involved, and their disruption may result in aberrant gene expression of *DLX5*, *DLX 6* and *DSS1*. It is also possible that position effects play a role here (a situation in which the phenotype expressed by a gene is altered by changes in the position of the gene within the genome, often by translocation). Especially *DSS1*, which is closely surrounded by translocation breakpoints that do not interrupt any known genes, may be susceptible to this position effect.[17]

1.3 SHFM2 (MIM 313350)

Only one family has been described where isolated SHFM is obviously transmitted as an X-chromosomal trait. In this Pakistani inbred kindred, the full manifestation of the SHFM phenotype was present in 33 males and 3 females. The males had monodactylous or bidactylous hands with bidactylous feet, the females were either normal or presented with mild deformities of the hands and/or feet. X-chromosomal inheritance was confirmed by linkage analysis, mapping the disease locus to chromosome Xq26-q26.1.[19]

1.4 SHFM3 (MIM 600095)

The third SHFM locus was initially mapped to a large interval at chromosome 10q24-25.[41,42] Studies of the naturally occurring *Dactylaplasia* (Dac) mouse were crucial to further investigate this locus.

The *Dac* mouse displays a phenotype that resembles the SHFM phenotype in humans. The phenotype of heterozygous mice consists of absent central digits, underdevelopment or absence of metacarpal/metatarsal bones and syndactyly. The homozygous mice display severe monodactyly. The SHFM3 locus at chromosome 10q24 is syntenic to the Dac region on mouse chromosome 19, thus making the *Dac* mouse an animal model for SHFM3.[27]

Two *dactylaplasia* alleles have been found in mice: Dac^{1j} and Dac^{2j}. Dac^{1j} is associated with an insertion 14 kb upstream of the *dactylin* gene. In Dac^{2j} the *dactylin* gene contains a 5.5 kb intronic insertion.[27] Thus, two different mutation events at different positions in and near the *dactylin* gene cause the Dac phenotype. *Dac* expression is further modulated by a modifier gene, *mdac*. To express the heterozygous phenotype, the animals must be homozygous for the *mdac* gene.[26]

It has been shown that the loss of central digital rays in affected limbs is caused by a defect in maintenance of the AER activity caused by a disruption of the *dactylin* gene. The *Dactylin* gene is a member of the *F-Box/WD40* gene family. These genes encode adapters that target specific proteins for ubiquitin mediated destruction. It has been proposed that a suppressor of AER cell proliferation exists, which is being regulated by *dactylin*. Normally, *dactylin* would mediate degradation of the suppressor, thereby allowing appropriate cell proliferation in the AER. In *Dac* mutants, the suppressor is not degraded, leading to decreased cell proliferation and premature elimination of the AER.[27]

The *DACTYLIN* gene in humans (*FBXW4* gene) has been mapped to chromosome 10q24 and has been shown to be 87% identical to mouse *dactylin* at nucleotide level.[43] Although the *DACTYLIN* gene seems to be the perfect candidate gene for SHFM3, until now no mutations have been detected in sporadic cases as well as in families which had been mapped to the SHFM3 locus.[28,35]

In an analysis of seven SHFM families, linked to the SHFM3 locus, a tandem duplication of approximately 500 kb was found at chromosome 10q24. The size of the duplication varied, with the minimal duplicated region being 440 kb and comprising the genes *LBX1*, *BTRC*, *POLL* and a disrupted copy of the *dactylin* gene.[28] Further mapping of the breakpoints of SHFM3 cases narrows the minimal duplicated region down to ~325 kb including only two genes, *BRTC* and *POLL*.[50]

In mice, the presence of a duplication in the *Dac* region in both alleles has been excluded. This confirms that the mutation mechanism is different than that of human patients with

SHFM. It is proposed that neither the *Dac* mutation in mouse nor the SHFM3 duplication in humans in itself causes the SHFM phenotype, but that these conditions might lead to complex alterations of gene regulation that would impair limb morphogenesis.[20] There are several genes within and near the duplication that are good candidates for involvement in limb malformation. For example, Lyle et al.[50] have shown that *BTRC* and *SUFU* are overexpressed in SHFM3. They regulate ß-catenin signalling, indicating that the ß-catenin signalling pathway is disrupted in SHFM3. Another gene that lies in the SHFM3 region is *FGF8*. Absence of *FGF8* in limb buds results in hypoplasia or aplasia of distal skeletal elements. In the *Dac* mouse, reduced *Fgf8* expression in the AER has been shown, but it remains unclear if this is a direct effect of the mutation or due to alterations in upstream pathways.[50,26]

Basel et al.[51] found a two-fold decrease in *DACTYLIN* gene transcript in five individuals with non-familial SHFM as compared to unaffected controls. *LBX1* gene transcript was also decreased, and *BTRC* and *FGF8* were low in patients as well as controls. Because of the proximity of these genes, it is possible that position effects cause the reduced expression levels. The fact that no mutations were detected in candidate genes in the SHFM3 region strengthens this hypothesis.[51]

1.5 SHFM4 (MIM 605289) - MUTATIONS IN THE *TP63* GENE

EEC syndrome (MIM 129900) is an autosomal dominant disorder characterized by ectrodactyly, ectodermal dysplasia and cleft lip/palate. In 1999, it was discovered that mutations in the *TP63* (tumour protein 63) gene at chromosome 3q27 are the cause of EEC syndrome.[46] Several EEC families were mapped to region 3q27 where previously an EEC-like disorder, limb-mammary syndrome (LMS, MIM 603543), had been mapped. Analysis of the *TP63* revealed heterozygous mutations in 9 unrelated families. The *TP63* gene is a homologue of the tumour suppressor gene *p53* and is also called *p63*.[46] *TP63* plays a critical role in the formation and differentiation of the AER. In studies with *TP63* knockout mice, these mice exhibited absent or truncated legs, abnormal skin, and lacked hair follicles, teeth, and mammary glands. In *TP63*$^{-/-}$ mice, the AER is absent during embryogenesis. This is likely to be due to a defect in ectodermal-mesenchymal signalling, which also causes the ectodermal dysplasia seen in these animals.[47]

Because of the phenotypical overlap between EEC syndrome and SHFM, two families with non-syndromic SHFM were screened for mutations in *TP63*. Linkage to all other known

SHFM loci was previously excluded. Two missense mutations were found in exons 5 and 7 of the *TP63* gene, thus adding *TP63* mutations as a cause of SHFM (SHFM4).[21]

In a large study with multiple families with EEC syndrome, isolated SHFM, and LMS, different mutations were detected between the syndromes, suggesting a phenotype-genotype correlation. *TP63* mutations were found in almost all EEC patients (40/43), but only in a small proportion of isolated SHFM patients (4/35) and in two of the three families with LMS. In EEC syndrome, mutations change specific amino acids in the DNA-binding domain (exons 5-8). LMS patients tend to have frame shift mutations near the 3' end of the gene. In contrast, in SHFM patients, nonsense and splice-site mutations as well as missense mutations are described. The fact that one mutation at codon 280 can cause EEC syndrome as well as SHFM, suggests that modifying genes play a role here.[48]

Recently, *TP63* and *DLX5/DLX6* were shown to co-localize in the embryonic AER. Also, $\Delta Np63\alpha$, the predominant *TP63*-form expressed in developing limbs, can activate *Dlx5* and *Dlx6* transcription. This transcription is disturbed by EEC and SHFM4 mutations, but not by AEC mutations (Ankyloblepharon-ectodermal defects-cleft lip/palate syndrome [MIM 106260], a syndrome also caused by *TP63* mutations). Together with the fact that $TP63^{EEC}$ combined with incomplete loss of *DLX5/DLX6* alleles results in aggravated limb phenotypes, these findings indicate that *TP63* lies genetically upstream of the *DLX* genes during limb development. The precise mechanism of this pathway is still incompletely known.[49]

1.6 SHFM 5 (MIM 606708)

Phenotypically, SHFM5 patients exhibit significantly more craniofacial abnormalities, mental retardation, and camptodactyly of the fingers.[2] Several patients have been reported with an interstitial deletion of chromosome 2q24.2-q31.1. Doles et al.[29] describe a boy with multiple abnormalities, including bilateral split foot. Four other cases with this deletion show several digital abnormalities of the hands and feet, including a wide gap between the first and second toe, wide halluces, brachysyndactyly of the toes, and camptodactyly of the fingers, associated with other abnormalities (mental retardation, craniofacial abnormalities). 2q31.1 seems to be the common deleted segment in these cases, but the phenotype seems to consist of a spectrum ranging from milder to more severe forms of limb abnormalities.

Del Campo et al.[22] report on two boys with several abnormalities consisting of a single bone in the zeugopod, monodactyly of the hands, mono- or bidactyly of the feet, and genital abnormalities. One patient has an interstitial deletion on chromosome 2q24.2-q31; while in the other patient 2q31.1-q32.2 is deleted. These phenotypes would fit well into the severe side of the SHFM spectrum. Since in the latter deletions the entire *HOXD*-cluster is removed, it was proposed that the limb and genital abnormalities are due to haploinsufficiency of the 5' *HOXD* genes (*HOXD9-HOXD13*).[22] These genes are critical for limb and genital tract development. However, it has been shown that a deletion of 2q31 removing the entire *HOXD*-cluster as well as the *EVX2* gene (upstream of *HOXD13*) causes synpolydactyly[44], and, for this reason, is unlikely to cause the SHFM phenotype.

Goodman et al.[44] suggest that the SHFM5 locus lies centromeric to the 5' *HOXD* cluster, between the *EVX2* gene and microsatellite marker D2S294 at 2q31. This interval comprises ~5 Mb, and candidate genes in this region are *DLX1* and *DLX2*. *DLX* genes are expressed in developing neuronal tissue. *DLX5* and *DLX6*, the candidate genes in SHFM1, are also expressed in differentiating osteoblasts, as discussed before. In mice, *Dlx1* and *Dlx2* are expressed in the AER and the PZ.[23] Interestingly, *Dlx1* and *Dlx2* knockout mice exhibit craniofacial abnormalities, however, no limb defects were observed in homozygous or heterozygous *Dlx1⁻/2⁻* mice.[45]

1.7 EVIDENCE FOR TWO NEW SHFM LOCI

Evidence for an additional SHFM locus was found in an affected family with non-syndromic ectrodactyly. Affected members had upper limb abnormalities without involvement of the feet. Linkage to the 5 known loci was excluded, and using a genome-wide scan this family was mapped to a 21 Mb interval on chromosome 8q. Mutation analysis of two candidate genes (*GDF6* and *FZD6*) in this region did not show any mutations in affected members of the family.[39]

In a large consanguineous family with autosomal recessive SHFM a further novel SHFM locus was identified at chromosome 12q13.11-13. In this family, all affected members except one had central feet reductions with or without hand involvement. One member had unilateral hand syndactyly with no feet involvement and is considered atypical. Linkage to other known loci was excluded. A screening of candidate genes revealed a homozygous missense mutation in the gene *WNT10b* in all individuals as well as in an asymptomatic female. *Wnt10b* acts as a key signalling molecule promoting osteoblastogenesis and inhibiting adipogenesis. The family member with atypical SHFM did not carry the mutation. The authors propose that either a second locus contributes to the manifestation of the SHFM phenotype or a suppressor locus prevented trait manifestation in the non-penetrant family member. Additionally, mutation analysis of *TP63* identified a rare insertion polymorphism in almost all affected individuals.[40]

1.8 SPLIT HAND/FOOT MALFORMATION AND LONG BONE DEFICIENCY

The split hand/foot malformation with long bone deficiency (SHFLD1; MIM 119100) should be considered a separate entity. The phenotype shows aplasia or hypoplasia of the tibia with associated split hand/split foot deformity, but is highly variable and can range from hypoplastic big toes to tetramonodactyly. Deformities of other long bones can also occur, including hypoplasia or bifurcation of the femur, hypo- or aplasia of the ulna as well as minor abnormalities of the patella or cup-shaped ears.[31] SHFLD1 is inherited in an autosomal dominant manner with reduced penetrance, although some authors found evidence for an autosomal recessive inheritance pattern. Recently, Babbs et al.[32] identified a locus for SHFLD1. In a patient with tibia hypoplasia, patella dislocation, fibular campomelia, and ectrodactyly a *de novo* chromosomal translocation was detected t(2;18)(q14.2;p11.2). The 2q14.2 breakpoint coincides with the homologous region of the ectrodactylous *Dominant hemimelia* mouse mutation and is proposed to represent a novel locus for SHFLD1.

A condition that shares several characteristics with SHFLD1 is *fibular aplasia with ectrodactyly*. Both conditions appear to be inherited in an autosomal dominant fashion, and express phenotypical variability and reduced penetrance. In fibular aplasia with ectrodactyly, there seems to be a lower penetrance rate in women. A multifactorial threshold model, with a major autosomal predisposing gene but a complex pattern of inheritance, could explain this.[33]

1.9 GENOTYPE/PHENOTYPE CORRELATION

Elliott et al.[1] have shown that in SHFM, a significant variable in discriminating genetic loci is preaxial involvement of the upper extremities. This was most frequently seen in patients with the SHFM3 genotype. Proximally placed thumbs and triphalangeal thumbs were observed to be specific preaxial variables associated with the SHFM3 locus. Of the 47 SHFM3 patients analyzed, 15 (31.9%) had a triphalangeal thumb. Preaxial involvement of the upper limb and, especially, triphalangeal thumb, are also associated with central clefting of the feet in SHFM3 patients. Thus, unmapped patients with TPT/split foot are likely to represent SHFM3 cases and should be evaluated for genomic rearrangements at 10q24. TPT may be identified only by radiographic analysis, emphasizing the importance of imaging these patients and their family members.[30]

In another descriptive epidemiological study, genotype-phenotype correlations were analyzed at a chromosomal level and several significant clinical features were found. Mental retardation in SHFM patients is closely related to karyotypic abnormalities, for example SHFM3 patients with a trisomy at 10q24 or SHFM5 patients with a deletion at 2q31. The candidate *DLX* genes in SHFM1 and SHFM5 are also expressed in the developing brain. When they are disturbed in their expression it is possible that they affect limb and brain development through a common pathway. Ectodermal involvement was, not surprisingly, most striking in SHFM4 patients. However, some EEC syndrome patients have also been mapped to the SHFM1 locus, but they did not express lacrimal involvement. Skin freckling is associated with the 3q locus rather than the 7q locus. Cleft lip/palate was reported at all SHFM loci, and most frequent in SHFM4 and SHFM5 patients. Hearing loss is a significant clinical feature; it is associated with SHFM1. The SHFM2 cohort, which was not included in the study, did not show any associated malformations or ectodermal involvement.[2]

2. HYPOTHESES

So far, estimations of the frequency of the five known SHFM loci have not been made based exclusively on phenotype. All previous publications included families, who had been mapped to a certain SHFM locus, as well as sporadic cases [34], or have focused only on one specific locus, i.e. SHFM3 [35].

In this study, we propose that the frequency of SHFM3 based on the typical phenotype is substantially high. In the clinical practice, non-syndromic cases of SHFM should not only be tested for mutations in the *TP63* gene which is the common diagnostic procedure at the moment, but primarily for genomic aberrations at the SHFM3 locus.

Furthermore, we suggest that there is no correlation between the size of the tandem duplication at the SHFM3 locus and the phenotype.

3. MATERIALS AND METHODS

3.1 PATIENTS

Patients were selected based on the presence of SHFM in one or more limbs. We included sporadic as well as familial cases. Informed consent was obtained from all patients. Clinical details are summarized in Table 1. None of the families had been previously mapped to a specific genetic locus. All patients were previously tested for mutations in the *TP63* gene by sequence analysis. Those who were tested positive were excluded from Multiplex Ligation-dependent Probe Amplification (MLPA) and quantitative PCR (qPCR) testing. Patient's DNA was extracted from blood leukocytes.

A positive control for the SHFM3 locus was obtained by array CGH testing, which revealed a duplication at locus 10q24 in a family with SHFM (index patient: 536, Fig. 2).

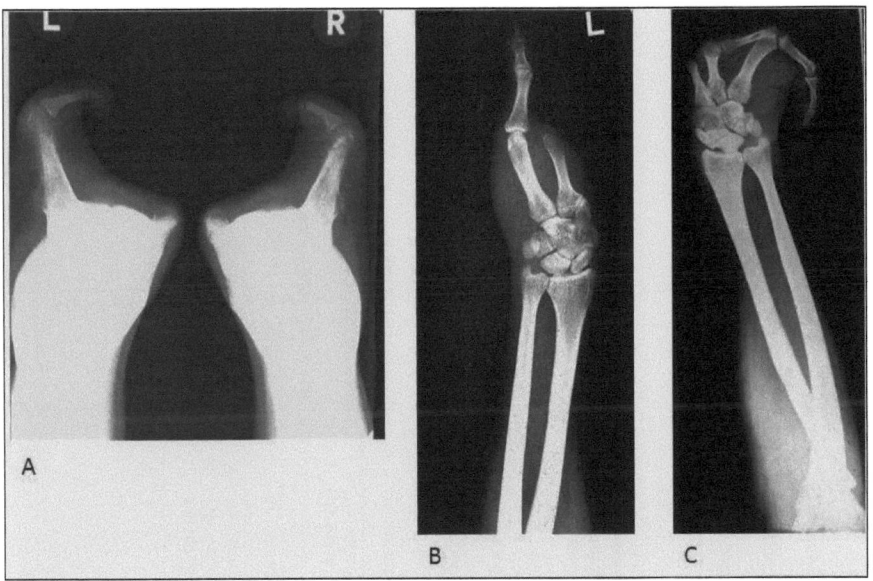

Figure 2. X-rays of the positive control case (patient 536). The patient exhibits monodactyly of the feet (A), monodactyly of the hands with absence or hypoplasia of metacarpals (B: left hand, C: right hand).

3. Materials and methods

Table 1. SHFM patients and their phenotypes.

Patient	Sex	Hand involvement	Feet involvement	Ectodermal involvement	Associated features	Familial/sporadic
536 (PC)	M	Bilateral monodactyly	Bilateral monodactyly	-	-	Familial
1391	M	Bilateral oligodactyly	Bilateral oligodactyly	Amelogenesis imperfecta, hair anomalies	-	Familial
4681	M	Left-sided oligodactyly, right-sided monodactyly	-	-	Partial right ulna-agenesia	Sporadic
4728	M	Right-sided syndactyly, bilateral oligodactyly	Toe hypoplasia, right-sided syndactyly	Ectodermal dysplasia	Short stature	Sporadic
4774	M	Bilateral oligodactyly	-	Oligodontia	-	Familial
5155	M	Left-sided oligodactyly	-	-	-	Unknown
5625	M	Bilateral oligodactyly	Bilateral oligodactyly	-	Uvula bifida, left-sided double kidney, vesico-uteral reflux	Sporadic
6126	M	Bilateral oligodactyly	Postaxial hexadactyly	Sparse, short hair, small teeth	Dysplastic cystic kidneys, mental retardation, anus imperforatus, microcephaly, micropenis, cleft palate	Unknown
6677	F	Bilateral monodactyly	Bilateral oligodactyly	-	-	Unknown
7151	M	Bilateral oligodactyly, right-sided syndactyly	Left-sided syndactyly	-	-	Sporadic
7174	M	-	Right-sided oligodactyly	-	-	Unknown
8198	M	Bilateral oligodactyly	-	-	-	Unknown
8436	F	Right-sided oligodactyly, left-sided syndactyly	Left-sided oligodactyly, right foot missing	-	Facial dysmorphisms, multiple stigmata	Unknown
8481	M	Bilateral oligodactyly	Bilateral oligodactyly	-	Bilateral fibula-aplasia, short stature	Unknown

3. Materials and methods

Patient	Sex	Hand involvement	Feet involvement	Ectodermal involvement	Associated features	Familial/sporadic
8529	M	Bilateral oligodactyly	Bilateral oligodactyly	-	-	Sporadic
8727	F	Bilateral oligodactyly	Bilateral oligodactyly	-	-	Unknown
8892	F	Ectrodactyly	-	Possible	-	Familial
9320	M	Bilateral oligodactyly	Bilateral oligodactyly	-	Cleft palate	Sporadic (Familial cleft palate)
9356	F	Bilateral oligodactyly	Bilateral oligodactyly	-	-	Unknown
9417	F	Bilateral oligodactyly	Bilateral oligodactyly	-	Small stature, mild mental retardation	Sporadic
9953	F	Left-sided syndactyly	Bilateral oligodactyly, syndactyly	-	-	Unknown
9995	M	Bilateral oligodactyly	Bilateral oligodactyly	Light hair, light skin	-	Unknown
A1384	M	Bilateral monodactyly	Bilateral monodactyly	-	Myopia	Familial
A1746	M	Bilateral oligodactyly, right-sided syndactyly	Left-sided oligodactyly, hypoplasia 5th digit	-	-	Sporadic
1473	F	Bilateral oligodactyly	-	-	Unilateral fibula-aplasia	Familial
2303	M	Bilateral oligodactyly	-	Mild ectodermal dysplasia	-	Sporadic
2269	F	Bilateral oligodactyly, unilateral syndactyly, unilateral preaxial polydactyly	Bilateral oligodactyly	-	-	Familial
2368	F	Bilateral monodactyly	Bilateral monodactyly	-	-	Sporadic
1202*	F	Bilateral oligodactyly	-	-	-	Unknown

F = female, M = male, - = not described

*Patient 1202 was not included in our original cohort of 28 patients and DNA was tested separately with array CGH.

3. Materials and methods

3.2 MULTIPLEX LIGATION-DEPENDENT PROBE AMPLIFICATION (MLPA)

All DNA samples were investigated with Multiplex Ligation-dependent Probe Amplification (MLPA) which allows detection of copy number changes at specific loci. With this technique unique probes of different length are added to a DNA sample. These probes consist of two separate oligonucleotides, each containing the universal PCR primer sequence (Fig. 3A). These parts are then hybridised to denatured sample DNA (Fig. 3B). This is followed by a ligation between the two probes (Fig. 3C); all ligated probes have identical end sequences, permitting simultaneous PCR amplification using only one primer pair (universal primer) (Fig. 3D). With this method up to 45 nucleic acid sequences can be amplified in one reaction.[14] Amplified probes are separated by electrophoresis and can be analyzed using MLPA analysis software. Relative amounts of amplification products, as compared to a control DNA sample, reflect the relative copy number of target sequences in the test DNA sample.

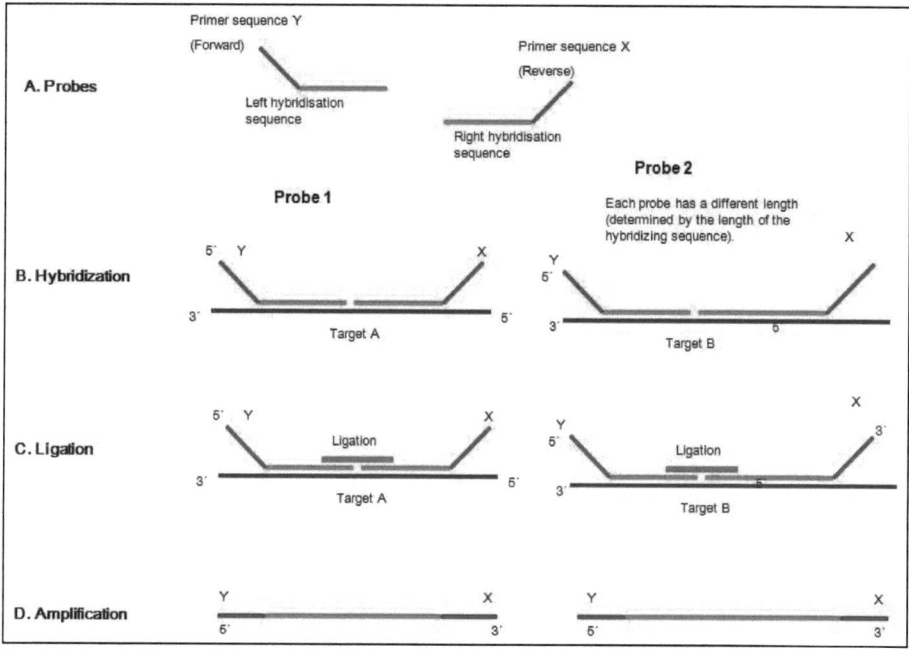

3. Materials and methods

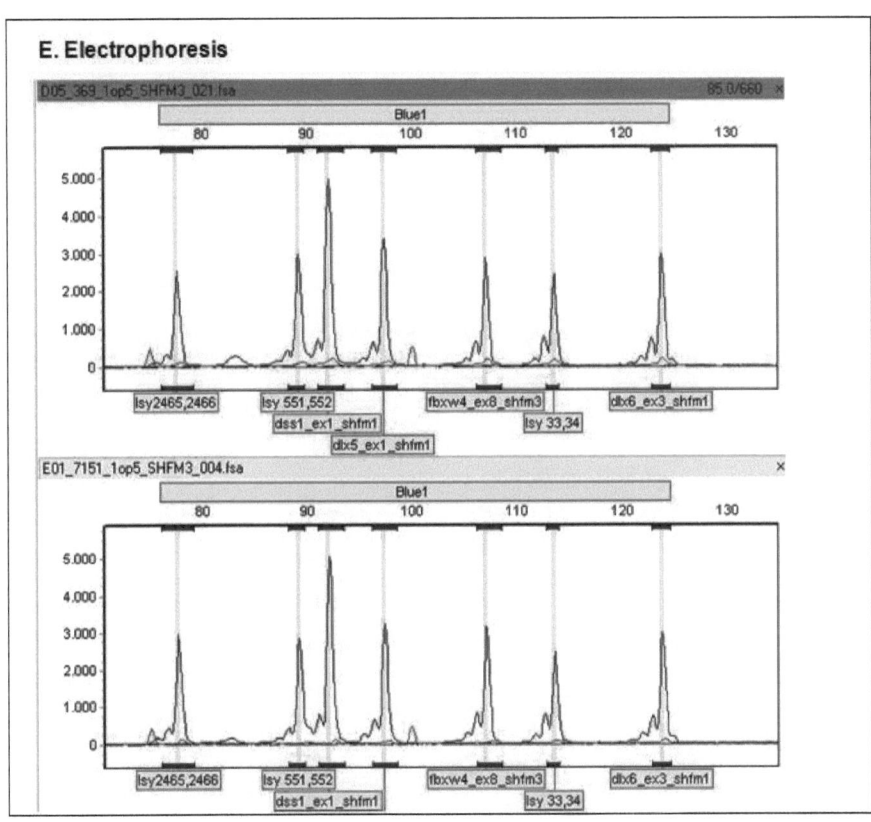

Figure 3. Principles of MLPA. Probes of different lengths (60-180 bp), consisting of two separate oligonucleotides (left and right hybridization sequence) contain a universal primer sequence that is fluorescently labelled (A). These probes are hybridized to denatured target DNA (B). The left and right hybridizing sequence are then ligated to each other (C). With the use of only one universal primer pair all probes are amplified in a single PCR reaction (D). Each amplification product has a unique length. The amplified probes are separated by electrophoresis and can be analyzed using MLPA analysis software (E). Modified from: http://www.mrc-holland.nl.

Probes were designed using the MRC-Holland Protocol (http://www.mrc-holland.com). Probe sequences were then blasted with UCSC Blat (http://genome.ucsc.edu) to certify that there are no homologies with other human sequences. A set of probes was designed for all the known SHFM-loci, which also included several control probes from other chromosomal loci. A second set was made with different probes for SHFM1 and SHFM3

3. Materials and methods

and control probes. After testing both sets with our positive controls, optimal probes were selected for the final probe set that was used in our experiments.

Probe sequences can be requested from E. Aten, Leiden University Medical Center, The Netherlands. A positive control for the SHFM1 locus was also provided by E. Aten.

For the MLPA reaction I used The MRC-Holland SALSA® MLPA® Kit and a PTC-100® Peltier Thermal Cycler.

1) Denaturation and hybridization of probes

DNA samples were diluted with TE buffer (10 mM Tris-Cl, pH 7.5; 1 mM EDTA) to a 5 µl solution of 40 ng/µl. 1.5 µl Probe-Mix and 1.5 µl Yellow Buffer was added to the sample and mixed. The sample was then denatured for 1 minute at 95°C in the thermal cycler followed by 16 hours at 60 °C to allow probe hybridization.

2) Ligation reaction

Temperature is first reduced to 54°C. Then 32 µl of Ligase-65 mix is added to the sample followed by an incubation step for 15 minutes at 54°C. The ligation is stopped by heating the samples to 98°C for 5 minutes.

    Ligase-65 mix:
    3 µl Ligase-54 Buffer A
    3 µl Ligase-65 Buffer B
    1 µl Ligase 65
    25 µl H2O

3) PCR reaction

5 µl of the ligation product is added to 20 µl PCR mix.

    PCR mix:
    2 µl Salsa PCR buffer Red
    1 µl Salsa PCR buffer Blue
    0.25 µl MLPA Forward primer Fam 20 µM
    0.25 µl MLPA Reverse primer 20 µM
    0.5 µl dNTPs 10 mM
    0.25 µl Salsa Polymerase
    15.75 µl H2O

    PCR program:
    5 min 94°C

3. Materials and methods

33 cycles: 20 sec 95°C
            30 sec 55°C
            1 min 72°C
20 min 72°C
Hold 4°C

4) Capillary electrophoresis:

The samples were analyzed using the ABI 3730 automated DNA sequencer (Applied Biosystems). First the PCR products were diluted 1:5 with formamide mix. Formamide mix (1 ml formamide + 6 μl ROX 500 Size Standard) is pipetted into the wells (10 μl per well). Then 1.5 μl of the diluted PCR product is added to the well. After this samples are denaturized in a thermocycler at 94°C for 3 minutes before capillary electrophoresis.

The data were analyzed with Genemarker Software (http://www.softgenetics.com). A template is created for each probe set. Probes are labelled according to length in the panel editor and this template is then used to analyze the data from the sequencing analysis. Peak detection threshold can be altered prior to analysis, depending on the peak intensity of the analyzed samples. Analysis settings are shown in Figure 4.

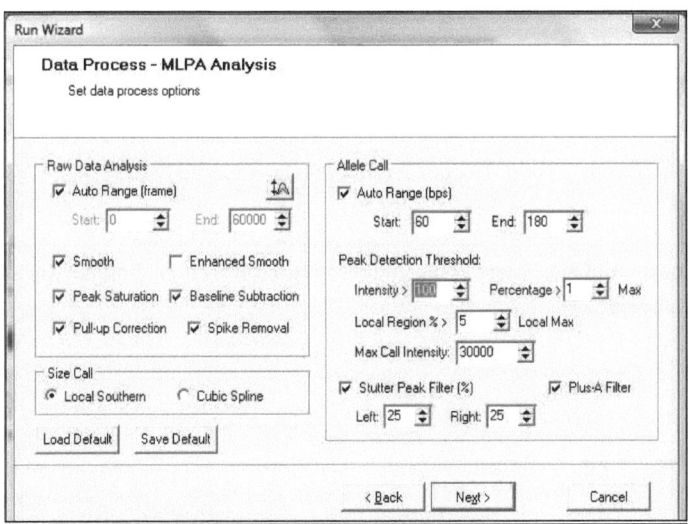

Figure 4. Peak detection settings in Genemarker software.

The software is designed to detect copy number changes in genomic sequences. In order to identify these copy number changes, it uses the peak intensity ratios of the sample DNA compared to control DNAs. The threshold for duplications and/or deletions can be defined by the user. Data points which lie outside of the threshold indicate a duplication or deletion. In our case, the threshold for duplications was set to ≥ 1.25 and for deletions ≤ 0.75.

3.3 REAL-TIME QUANTITATIVE PCR SHFM 3 LOCUS (10Q24)

Initially we used five amplicons (P1, P3, D1, D3, and D13) that were located at the SHFM3 locus on chromosome 10q24 (Table 2).[28,50] After detecting a duplication that was larger than the ~440kb spanning region which included *POLL*, *LBX1* and *BRTC1*, we used additional amplicons (18, 19 and 20) to obtain an estimation of the breakpoint of these duplications (locations of primers are shown in Figure 5).

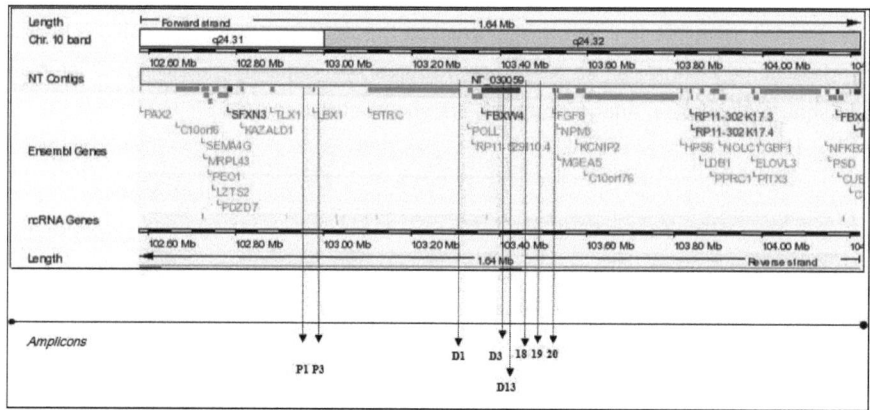

Figure 5. Location of amplicons at the 10q24 locus. Modified from: Ensembl Genome Browser (http://www.ensembl.org).

Table 2: Primer sequences and genomic positions on chromosome 10.

Primer	Sequence	Position start [bp]	End [bp]
P1 for	CACACACCACAGTCCAAAGG	102880148	102880167
P1 rev	TTCAACCAGATTGCATTCCA	102880208	102880227
P3 for	CCACCAACACCAAACCTCTT	102981795	102981814
P3 rev	GAGGCCAACCAATTAGGTCA	102981865	102981884
D1 for	AGCGGTGTTAACCATCACCT	103319122	103319141
D1 rev	TCATCCTGGTGCCTTTCACT	103319190	103319209
D3 for	CCATCTAAGGTCCTGCCTGA	103418716	103418735
D3 rev	GCCAGATGGATCAGTGACCT	103418656	103418675
D13 for	CACATCACCCTCCAGAAACA	103419366	103419385
D13 rev	GGGCAAAGATAGGATCAGCA	103419431	103419450
18 for	CCTACCAGTTCCGTCCAGATG	103423335	103423355
18 rev	GCCCAGCAAAGACTCCCAG	103423293	103423311
19 for	AGCTCATCGTGGAGACGGAC	103521294	103521313
19 rev	AGGCCCGTCTCGGCTC	103521250	103521265
20 for	TGCCAGAAACTTTCCTTGCTAAT	103537211	103537233
20 rev	GGCCACACTTGGGTCAGTTAC	103537154	103537174

To detect copy number changes at the SHFM5 locus we used four amplicons situated in and around the *DLX1* gene (Table 3).[44]

Table 3: Primer sequences and genomic positions on chromosome 2.

Primer	Sequence	Position start [bp]	End [bp]
DLX1 for	CCCAGGACGATTTATTCCAG	172659705	172659724
DLX1 rev	CTCTCCGGCAGAGCTAGGTA	172659768	172659787
D2S335 for	TGCCATCACAGGCTAGACAC	172267875	172267894
D2S335 rev	GCTTTAATTCGCCCTTACCA	172267937	172267956
HOXD10 for	TAGTCGTACCCTCCCACCAC	176682743	176682762
HOXD10 rev	ATTGCTTGGTCACATCGTCA	176682813	176682832
EVX2 for	CGGGTCCGATTTTGATATTG	176654288	176654307
EVX2 rev	CTTCCCCGTTTATGGAGACA	176654358	176654377

3. Materials and methods

Primers were selected using Primer 3 (http://primer3.sourceforge.net). Primer sequences were checked by BLASTN (http://www.ensembl.org) against the human genome to assure they were specific for the region under study. As a calibrator we used amplicons within the *Factor VIII* gene and *Albumin*.

All PCRs were performed using a qPCR mastermix (Power SYBR green, Applied Biosystems). Reactions were set up using 96-well-plates, in a 24 µl volume with three replicates per sample.

For every primer pair a premix was prepared consisting of:
    5 µl forward primer [100 pm/µl]
    5 µl reverse primer [100 pm/µl]
    190 µl H2O

2 µl of primer premix is added to each qPCR reaction well. Afterwards 12 µl of SYBR green mix (Power SYBR green, Applied Biosystems) is added. For the qPCR reaction 20ng of DNA are needed in every well. A 50 ng/µl predilution is prepared at first, which is then again diluted to a final concentration of 2 ng/µl. 10 µl of the reference or patient DNA dilution is added to the respective wells to reach a final amount of 20 ng per well. Each well contained 24µl reaction volume (2 µl primer mix, 12 µl SYBR green mix, 10 µl DNA). Reactions were run in a 7500 Real Time PCR System thermocycler (Applied Biosystems).

The following settings were used:
    95°C for 8 min
    95°C for 15 sec   40 cycles
    60°C for 35 sec   40 cycles

By intercalation of the SYBR green dye in the double-stranded amplification product the amplification process can be monitored cycle by cycle. The evaluation of the qPCR data was done by the comparative CT method. The CT value is the cycle number at which the amplification curve of a specific qPCR product crosses a fixed threshold. The threshold is set in the exponential phase in which the amplification occurs at a most accurate rate. The CT values of patient and reference DNA samples can be compared to each other. The value depends on the amount of starting DNA material. If there is a higher copy number the threshold will be reached faster and the CT value will be low. If there is less starting material in the sample DNA it takes more cycles to reach the threshold and the CT value will be higher (Fig. 6). Graphs were designed using Excel (Microsoft Cooperation).

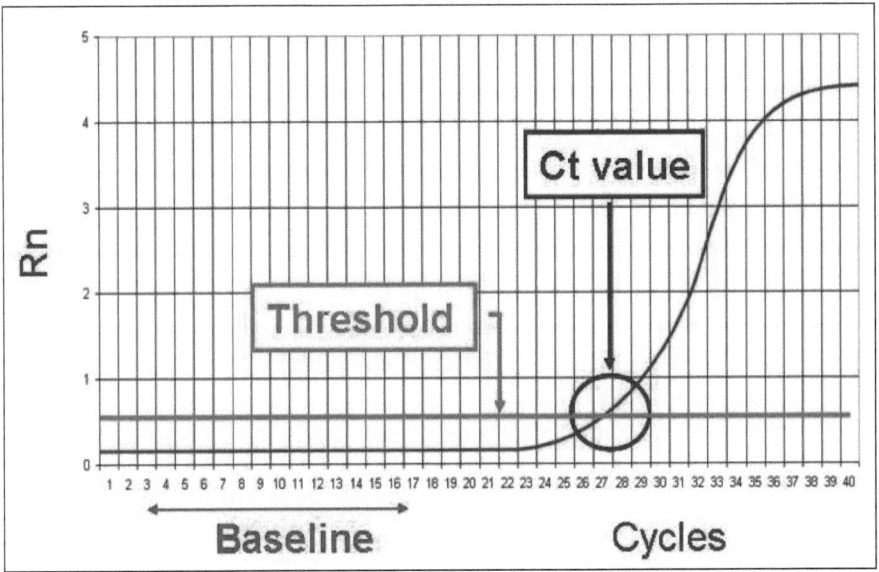

Figure 6. Comparative CT method. The CT value is measured at the point where the amplification curve (blue line) crosses the fixed threshold (in the exponential phase). Modified from: http://www.appliedbiosystems.com.

3.4 ARRAY CGH

Conventional comparative genomic hybridization (CGH) is a method developed to detect DNA copy number changes (chromosomal gains and losses) across the whole genome. Patient DNA and reference DNA of a healthy person are labelled differently with fluorophore-coupled nucleotides and then co-hybridized to normal metaphase chromosomes. DNA of the patient is labelled in green, reference DNA is labelled in red. The relative DNA copy number variation can be detected by measuring the fluorescence ratios along the length of the chromosomes. Chromosomal regions equally represented in both samples appear yellow and have a ratio of one; deleted regions are red and have a ratio below one, amplified regions appear green and have a ratio higher than one. CGH has been used mostly in tumour cytogenetics and to detect chromosomal abnormalities in genetic disorders. However, a disadvantage of this method is that the resolution is limited to 10-20 Mb.

In a newer method called array CGH (micro-array based CGH) the labelled DNA is hybridized onto an array of DNA spots printed on a glass slide, instead of being hybridized to metaphase chromosomes. Therefore, a BAC (bacterial artificial chromosome) clone library, cDNAs or oligonucleotide sequences that cover the entire genome are used as hybridization targets instead of metaphase chromosomes. This increases the resolution to 50 kb – 1 Mb depending on the array platform. Using oligonucleotide arrays is cheaper, faster, and offers a higher resolution than using BAC arrays, although a larger amount of DNA is needed (about 1 µg, compared to 200-400 ng for BAC arrays).[52] It is important to always match the sex of the reference DNA to the sex of the patient's DNA to detect chromosomal aberrations on sex chromosomes.[53] Advantages of array CGH are that this method does not require dividing cells (as in for example FISH or karyotyping) and that it enables analysis of the whole genome in a single experiment. However, array CGH (as conventional CGH) is unfortunately limited in its ability to detect mosaicism and is not able to detect balanced chromosomal translocations, inversions and whole-genome ploidy changes. Nowadays, array CGH is used for a variety of purposes, including searching for chromosomal imbalances in patients with mental retardation, congenital abnormalities, and tumour staging and prognosis.[52,53]

The following protocol was used for 244K oligonucleotide arrays (Agilent Technologies):

1) Restriction digest

1 µg of DNA was digested using a Digestion-Mix consisting of:

H$_2$O	2.0 µl
10x Buffer C	2.6 µl
BSA (10 µg/l)	0.2 µl
Alu I (10U/l)	0.5 µl
Rsa I (10U/l)	0.5 µl

The DNA and the digestion mix were incubated in a PTC-100® Peltier Thermal Cycler for 2h at 37°C. The reaction was then stopped by heating the cycler to 65°C for 20 minutes and thereby inactivating the restriction enzymes.

2) DNA labelling (random priming)

Digested DNA was prepared for labelling using 2.5x concentrated random primers from Invitrogen for 10 minutes at 99°C in the thermocycler.

Thereafter, DNA was labelled with:

Random Label dNTPs (Invitrogen)	7.5 µl
(dATP, dGTP, dTTP 1.2 mM each, dCTP 0.6 mM)	
Cy-3-dCTP or Cy5-dCTP 1 mM (Amersham)	2 µl
Klenow fragment, 40 U/µl (Invitrogen)	1.5 µl

in the thermocycler for 2h at 37°C. The reaction was stopped by adding 7.5 µl of Stop Buffer. Patient DNA was labelled with Cy3, reference DNA with Cy5. Labelled DNA probes were cleaned using Sephadex G-50 columns (Amersham) according to the manufacturer's protocol to remove unincorporated nucleotides. To confirm successful labelling, 3 µl of labelled DNA was loaded onto a 1% agarose gel with ethidium bromide and the rate of incorporation later measured using a NanoPhotometer (Implen GmbH). Values should lie between 25-40 pmol/µg for Cy3 and between 20-35 pmol/µg for Cy5.

3) Purification and precipitation

Patient and reference DNA were combined and Cot1-DNA was added to block repetitive sequences. The DNA was then precipitated by adding 3M NaAc (1/10 of the total volume) and 2.5 volumes of EtOH 100%, incubated for 30 minutes at -80°C and centrifuged at 13000 rpm for 30 minutes at 4°C. Pellets were washed with 500 µl 70% EtOH and centrifuged at 13000 rpm for 10 minutes at 4°C. Pellets were then dried at 37°C and dissolved with 24 µl H_2O. Thereafter 30 µl of pre-heated 2x Agilent Hybridization Buffer and 6 µl of 10x Agilent Blocking Agent were added and tubes shaken at 37°C. To prepare for hybridization, DNA was denaturized for 3 minutes at 95°C and then preannealed for 30 minutes at 37°C. Before loading the samples onto the oligonucleotide array, they were preheated to 65°C.

4) Array preparation and hybridization

The Advalytix Slide Booster (with 4 chambers, in each of which one array can be hybridized) was prepared with Advason coupling liquid AS100 put onto the impulse fields. 244k Agilent oligonucleotide arrays were used as arrays with 25x60 lifterslips (Implen). Slides were heated to 65°C and the pre-heated probe was added to the slide. Each slide was then incubated for 17h at 65°C to allow hybridization of the sample to the probes.

5) Array washing

Slides were rinsed with 2x SSPE buffer, consisting of:

175.3 g NaCl

27.6 g NaH$_2$PO$_4$ x H$_2$O
7.4 g EDTA
800 ml H$_2$O
NaOH to adjust to pH 7.4

They were then washed for 5 minutes in the dark with wash buffer 1 (0.5x SSPE, 0.025% SDS). Afterwards washed again for 2 minutes at 37°C with wash buffer 2 (0.1x SSPE, 0.025% SDS). Finally they were rinsed with 0.1x SSPE without SDS and dried by centrifugation at 900 rpm for 3 minutes.

6) Array scanning and analysis

Slides were scanned in an Axon 4100A laser scanner using GenePix Pro software (Axon Instruments). The image was then saved as a multi-image .tif file and analysed with Agilent Feature Extraction software. Gains and losses in copy number throughout the genome were detected and visualised by the CGH Analytics software (Agilent Technologies) using the ADM-2 algorithm. The aberrant regions must then be checked in i.e. Ensembl genome browser (http://www.ensembl.org) to identify pathological aberrations and exclude common copy number variants present in the normal population.

4. RESULTS

A total of 28 patients were initially tested for duplications or deletions at all known SHFM loci with MLPA. The SHFM2 locus was used as a sex control because of its position on the X chromosome. Unfortunately it was not possible to perform MLPA in subjects 8727, 5155, and 6677 because of the minor quality of the DNA. It was, however, possible to test these patients' DNA with qPCR for the SHFM3 locus.

Using MLPA, we found a duplication (ratio $\geq$ 1.25) at the SHFM3 locus in 6 patients. Experiments were repeated several times to confirm results. Summary of the results for SHFM1, SFHM2, SHFM3, SHFM4, and SHFM5 is shown in Figures 7-11. Three control probes were always included, but are not shown in the figures. We were able to confirm the duplication at the SHFM3 locus that we detected by MLPA in these patients, with quantitative PCR. Additionally, we found a duplication in SHFM3 in subject 8727, in which MLPA testing was not possible (Figures 12 and 13).

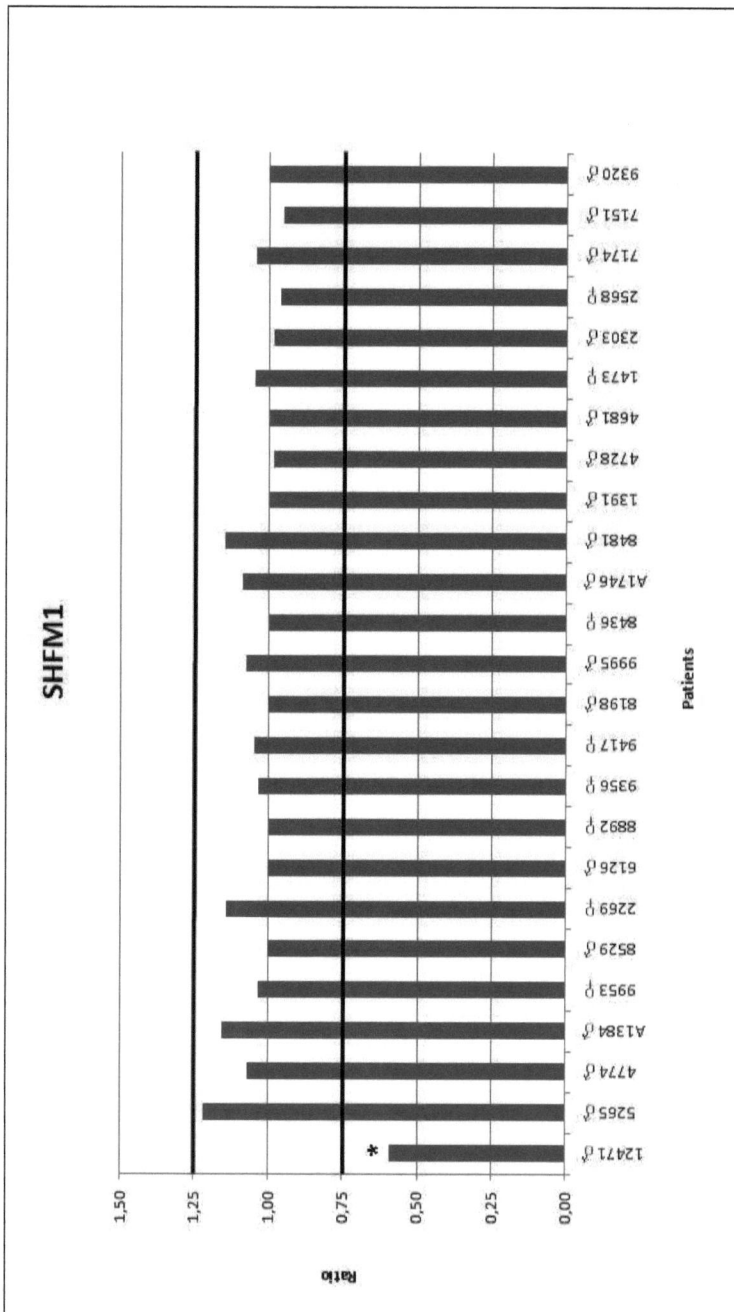

Figure 7. MLPA results for probe SHFM1. Patient 12471, marked by an asterix, shows a deletion at this locus (ratio ≤ 0.75) and is used as a positive control (courtesy of E. Aten, LUMC, the Netherlands). In the other patients SHFM1 is normal (2 copies).

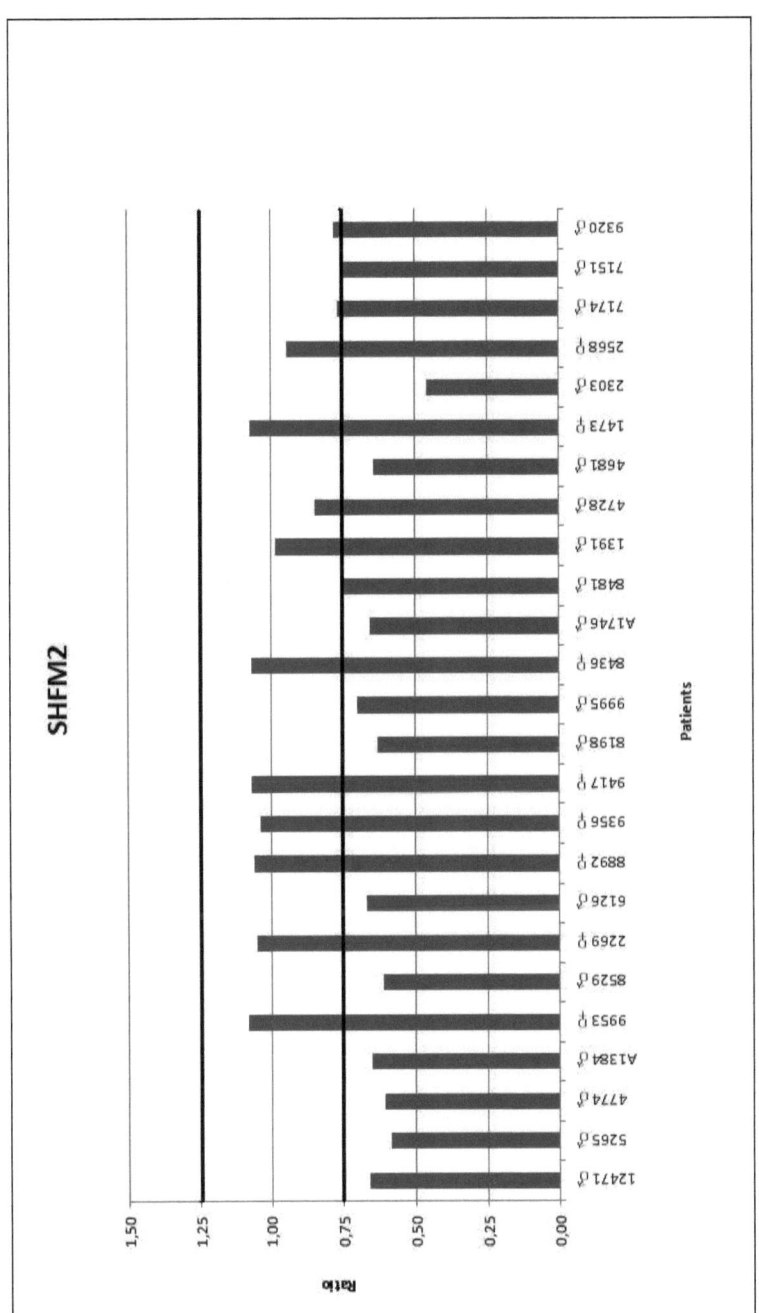

Figure 8. MLPA results for probe SHFM2. This probe was used as a sex control. Female patients show two copies of this locus on the X-chromosome (ratio ≥ 0.75 and ≤ 1.25). Male patients display one copy of the X-chromosome (ratio ≤ 0.75). See also Table 1 for a comparison of patient's gender.

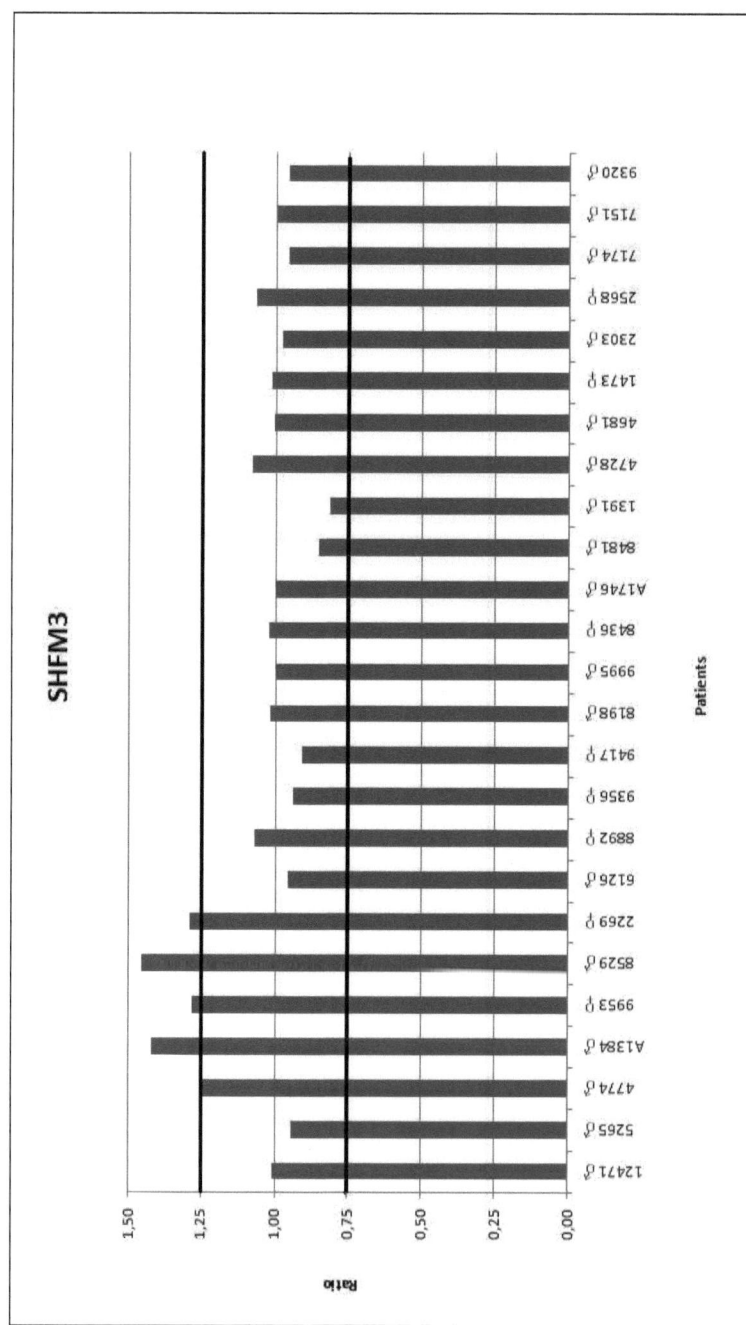

Figure 9. MLPA results for probe SHFM3. A duplication (ratio ≥ 1.25) is detected in patients 4774, A1384, 9953, 8529, and 2269.

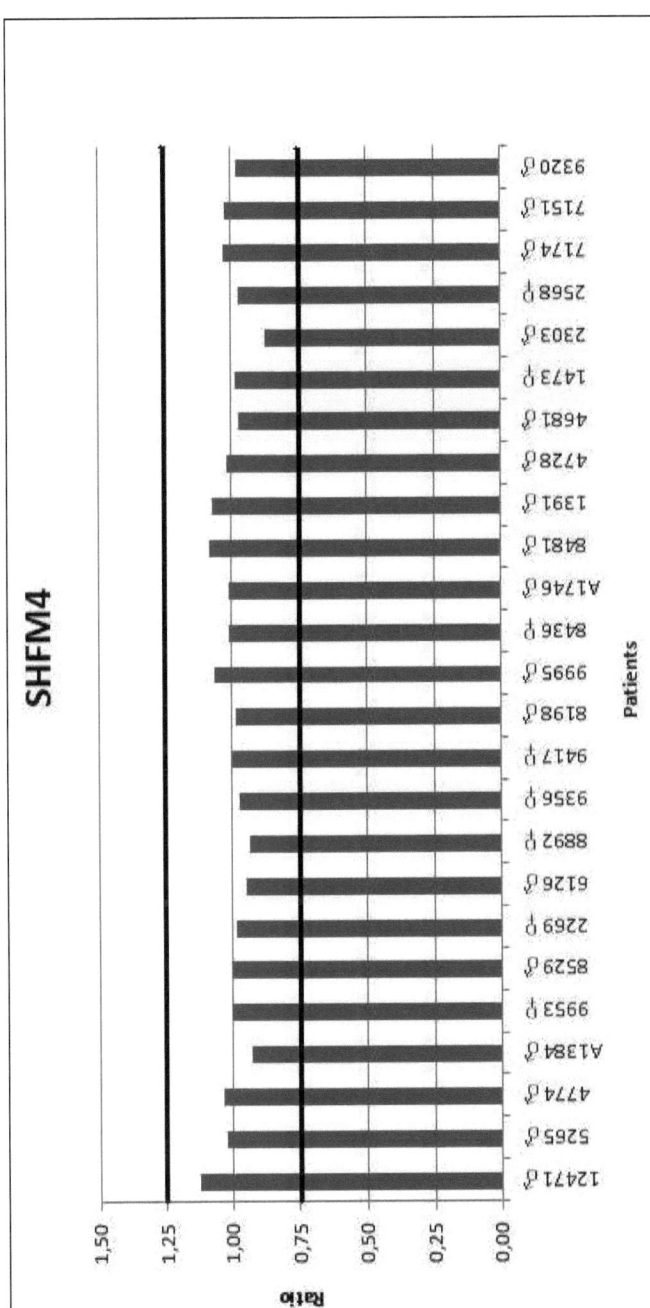

Figure 10. MLPA results for probe SHFM4. No copy number changes are observed at this locus in any of the patients.

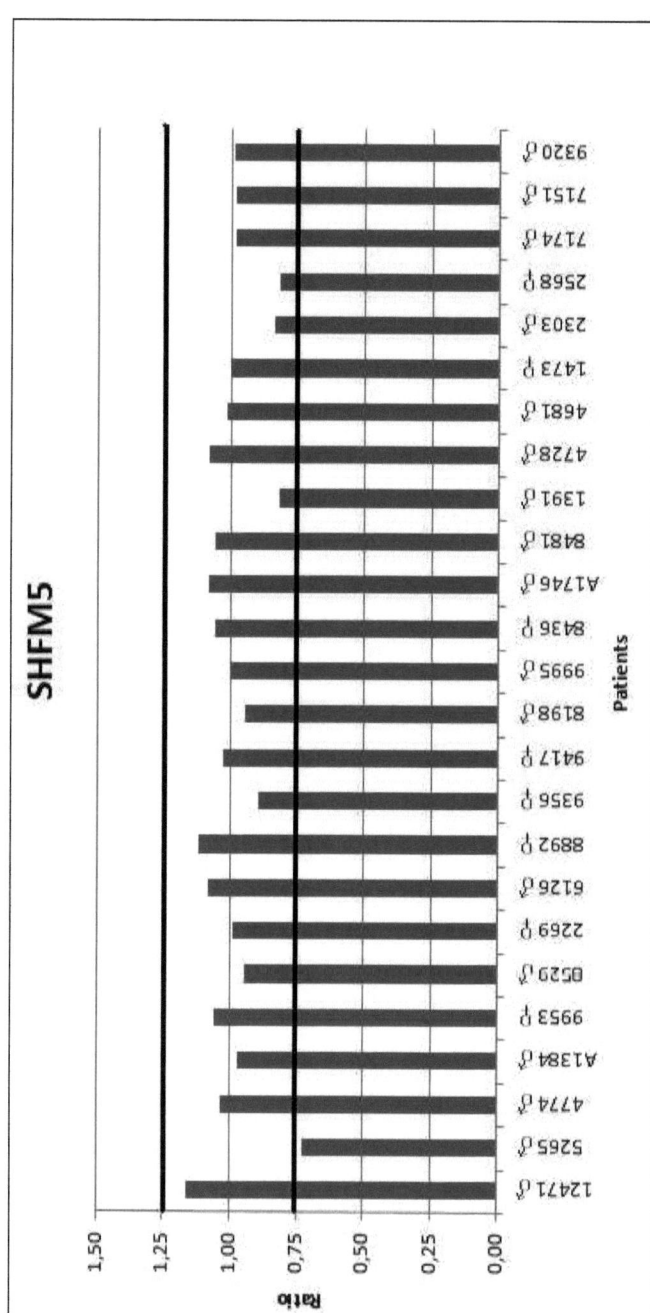

Figure 11. MLPA results for probe SHFM5. In patient 5265 there seems to be a deletion at this locus (ratio ≤ 0.75). In the other patients two copies of the SHFM5 locus exist

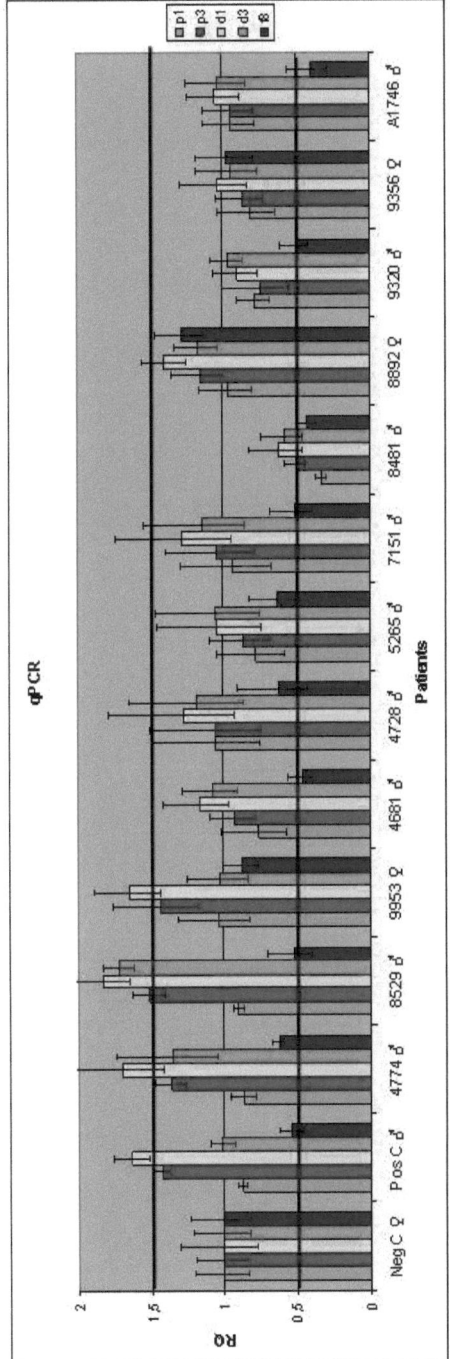

Figure 12. Quantitative PCR results with amplicons p1, p3, d1, d3 and f8 (SHFM3). A duplication (RQ ≥ 1.5) is seen in patients 4774, 8529, and 9953. Patient 4774 and 9953 show a larger duplication including amplicon d3; the duplication in the positive control (536) and patient 9953 includes amplicons p3 and d1.

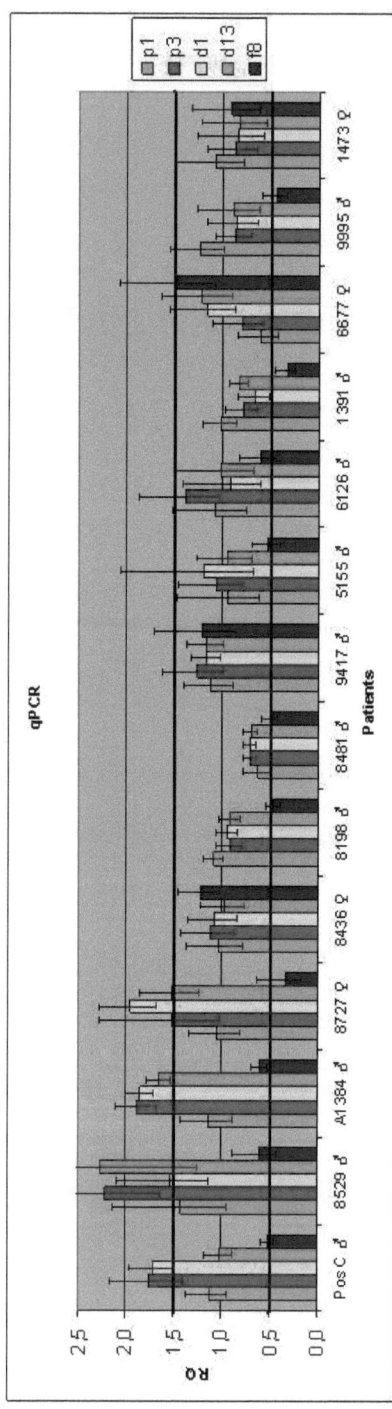

Figure 13. Quantitative PCR results with amplicons p1, p3, d1, d13 and f8 (SHFM3). In the positive control, p3 and d1 are duplicated (ratio ≥ 1.5). A larger duplication is seen in patients 8529, A1384, and 8727, in which amplicons p3, d1 and d13 show a duplication.

4. Results

The phenotype of the patients in which we detected the SHFM3 duplication was very diverse and includes one case (4774) of SHFM with ectodermal involvement. An overview of the phenotypes of SHFM3 positive patients is shown in Table 4. No copy number change was detected in any of the patients with associated syndromes.

Table 4: SHFM3 positive patients: size of duplication and phenotype.

Subject	Size of duplication	Phenotype
9953	~435 kb	Bilateral oligodactyly + syndactyly feet, syndactyly left hand
536	~435 kb	Monodactyly all four extremities
2269	~440 kb	Bilateral oligodactyly hands and feet, unilateral syndactyly and preaxial polydactyly hand
4774	~440 kb	Bilateral oligodactyly hands, oligodontia
8529	~440 kb	Bilateral oligodactyly hands and feet
8727	~440 kb	Bilateral oligodactyly hands and feet
A1384	~490 kb	Monodactyly all four extremities

Interestingly, we observed duplications of different sizes. In patient 536 and 9953 amplicons P3 and D1 are duplicated, but D3 and D13 are not. The terminal breakpoint lies between amplicon D1 and D3 and the duplication comprises about 435 kb. Patients A1384, 8529, 8727, and 4774 had a duplication that included P3, D1, D3, and D13, thus a slightly larger duplication of ~440 kb.

Further testing with additional amplicons revealed that in patient A1384 amplicon 18, but not 19 and 20 were duplicated, and in subject 8529 none of the additional amplicons were duplicated (Figure 14). Thus, the largest duplication we measured has a terminal breakpoint that lies between amplicon 18 and 19 (duplication size ~490 kb).

4. Results

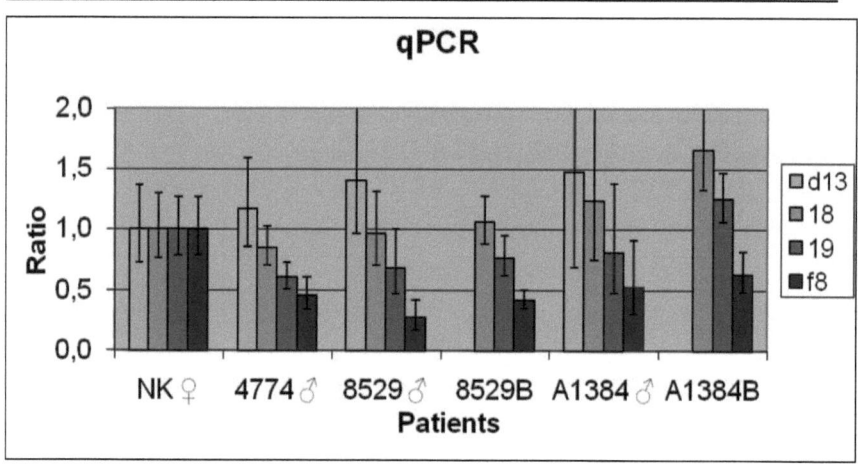

Figure 14. QPCR testing of patients 8529, 4774 and A1384 with amplicons D13, 18 and 19. Amplicon 20 was normal in all patients and is not shown here. qPCR was repeated in 8529 and A1384 (indicated with "B") because of large standard deviation in the first analysis. Patients 4774 and 8529 did not show a duplication of any of the additional amplicons, thus the breakpoint must lie between amplicons D13 and 18. In patient A1384, a duplication of amplicon 18 is not clear in the first analysis (large standard deviation), but clearly shown in the replication. This indicates that the breakpoint of this duplication is located between amplicon 18 and 19.

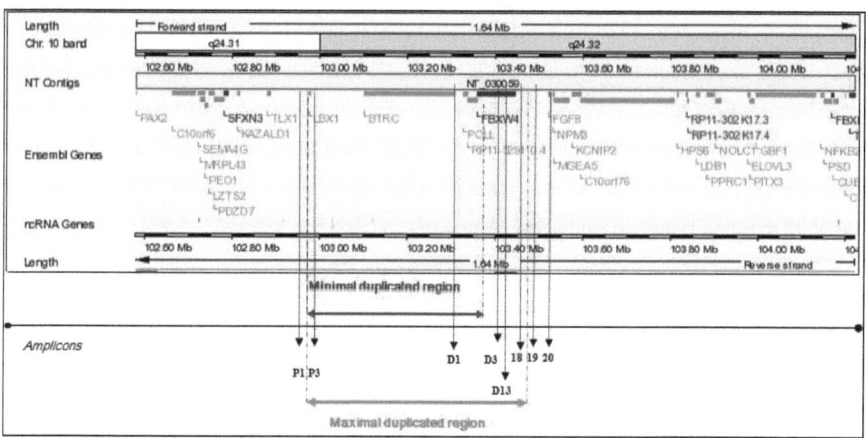

Figure 15. Minimal and maximal duplicated region detected in SHFM3 patients. The minimal duplicated region comprises approximately 435 kb (breakpoint between amplicon D1 and D3), the maximal duplicated region spans about 490 kb (breakpoint between amplicon 18 and 19). Modified from Ensembl Genome Browser (http://www.ensembl.org).

4. Results

In patient 5265, MLPA testing repeatedly showed a possible deletion at the SHFM5 locus (ratios varying from 0.778 to 0.725). Because we could not obtain a conclusive result, primers were designed for qPCR analysis at the SHFM5 locus. As shown in Figure 16, no deletion was detected with any of these amplicons.

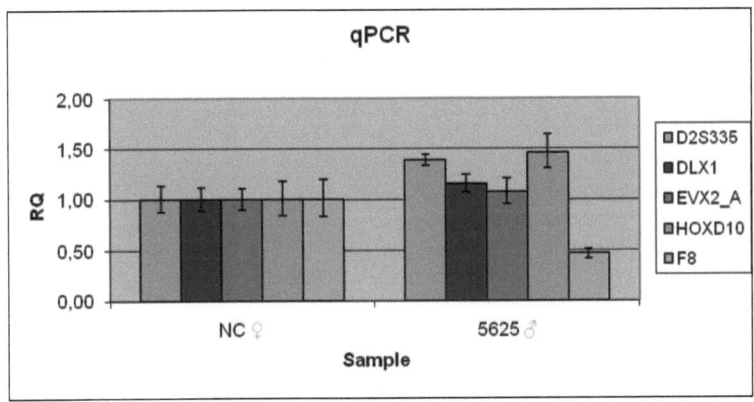

Figure 16. QPCR testing of subject 5625 with amplicons located at the SHFM5 locus. No deletion (ratio ≤ 0.5) was detected with any of the amplicons used. F8 is used as a sex control.

4.1 ARRAY CGH RESULTS

Because we could not confirm the possible deletion in patient 2565 at the SHFM5 locus with qPCR, we performed an array CGH with the patient's DNA to look for copy number variations, especially at chromosome 2q31. We did not detect any copy number variations in this region nor did we find any other pathologically relevant CNVs in the rest of the patient's genome with a resolution of ~50kb.

Patient 1202 is a *TP63* negative patient not included in our original cohort, on whom we only performed an array CGH to detect pathological copy number variations in the genome. Interestingly, we detected a duplication at chromosome 7q21.3 proximal to the SHFM1 critical region which comprises the candidate genes *SHFM1*, *DLX5* and *DLX6* (Fig. 17).

4. Results

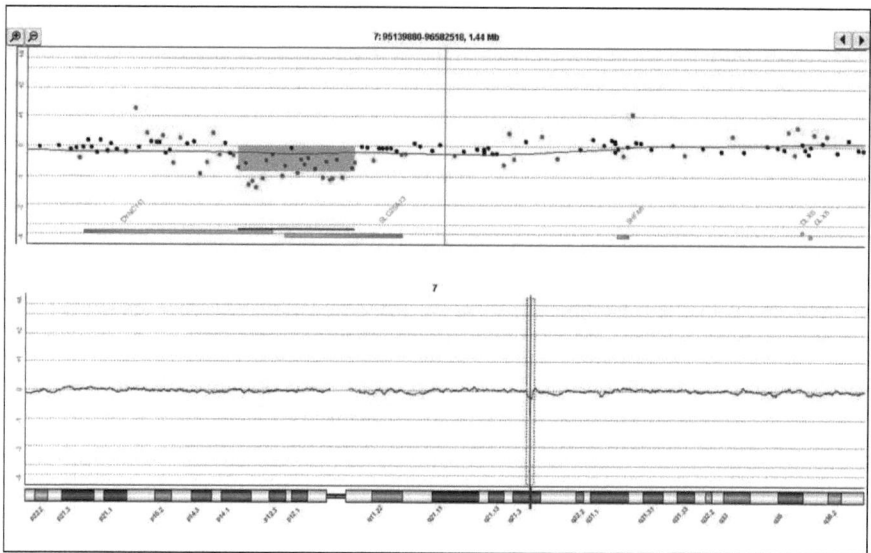

Figure 11. Array CGH result of patient 1202. An interstitial deletion of 1.44 Mb is detected at chromosome 7q21.3 proximal to the SHFM1 critical region which includes the candidate genes *SHFM1*, *DLX5* and *DLX6*. The upper panel shows the SHFM1 locus with the deleted region shaded in blue. Genes are indicated below the profile. Each dot represents one oligonucleotide probe on the array. A view of the whole chromosome 7 profile is shown in the lower panel with the aberrant region marked by a vertical blue line.

4. Results

4.2 FAMILIAL CASES

DNA of several family members of familial cases (patient 4774 and patient 2269) was available for testing. Pedigree of the family of patient 4774 is shown in figure 18.

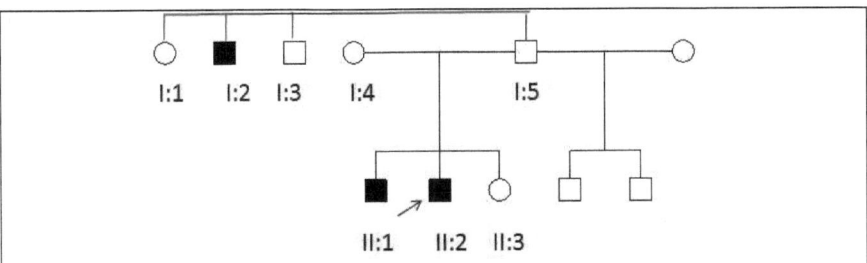

Figure 18. Pedigree of the family of patient 4774. II:2 is the index patient (4774) with bilateral oligodactyly of hands and feet and mild oligodontia. II:1 (patient 4775) exhibits bilateral split hand and oligodontia. The parents (I:4 and I:5) are asymptomatic. I:2 seems to have an unrelated limb malformation with digits 3-5 missing unilaterally. Family member I:3 presents with mental retardation.

Interestingly, we detected a duplication at SHFM3 in both patient 4774 and his brother (4775), but neither the father (4777) nor the mother (4776) carried the duplication (Figure 20). It is very likely that the duplication 10q24 is present as a germline mosaicism in one of the parents. The duplication was transmitted to the two affected sons but not the daughter. A brother of the father has a unilateral hand malformation, which seems to be caused by another condition and which phenotype is not typical for SHFM.

The second familial case concerns patient 2269, of whom the father (2305) is the only affected family member (for X-ray images see Fig. 19). The father, patient 2305, displays a much milder phenotype with syndactyly of the feet and no ectrodactyly. With qPCR testing, we found the same duplication as in patient 2269 (Figure 20). This is a good example of the intrafamilial variability of split hand/split foot malformation. It is unclear why amplicon d13 shows such a high relative quantification in both patients compared to amplicons p3 and d1 and amplicon d13 in the other patients with SHFM3 duplication. One possible explanation is a copy number of 4 copies i.e. a triplication compared to three copies of the other amplicons. This hypothesis was not investigated further in this study.

4. Results

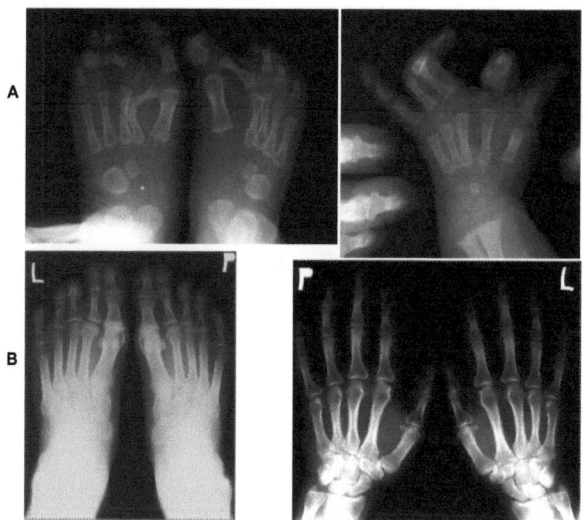

Figure 19. X-rays of patient 2269 and patient 2305. Patient 2269 exhibits bilateral split feet and unilateral left-sided split hand (A). She also has polydactyly of the left hand, which is not shown here. The patient's father (2305) exhibits cutaneous syndactyly of the feet and no hand deformities (B).

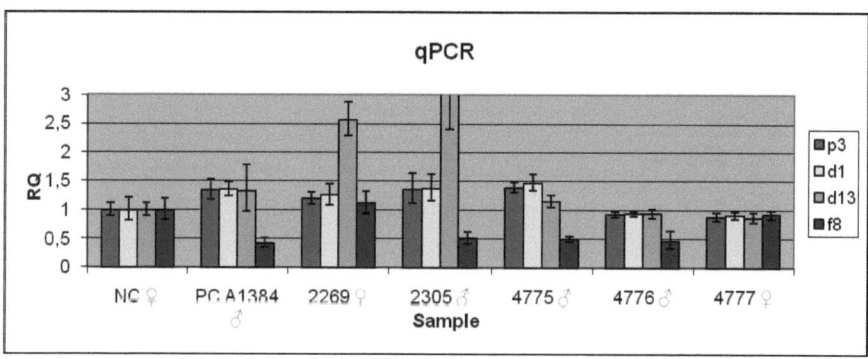

Figure 20. QPCR testing of family members of affected patients 2269 and 4774. The same duplication comprising amplicons p3, d1 and d13 (RQ ≥1.5) was detected in the father of patient 2269 (patient 2305). Amplicon d13 shows an increased RQ value in these two cases. The brother of patient 4774 (patient 4775) who is also affected carries the same duplication as the index patient. The parents (4776 and 4777) have normal copy numbers for all amplicons investigated.

As outlined above patients with a *TP63* mutation were excluded from testing by MLPA and qPCR. Of all 28 investigated patients, 7 patients harboured a duplication at the SHFM3 locus. No aberrations were found at the other SHFM loci in our cohort. Thus, the frequency of SHFM3 duplication in SHFM patients in our study, based only on SHFM phenotype, is 25% (Fig. 21). A duplication proximal to the SHFM1 locus was observed in patient 1202, but is not further discussed here because this patient was not part of our original cohort.

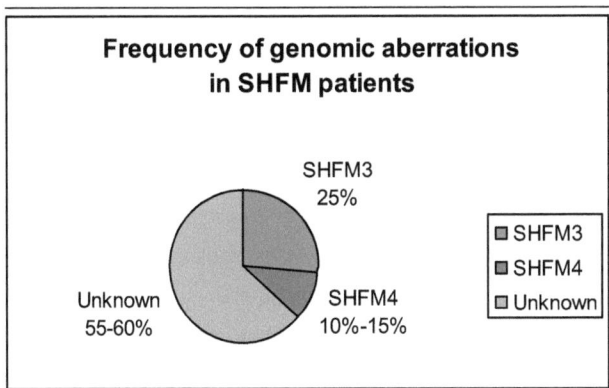

Figure 21. Frequency of genomic aberrations in SHFM patients. In our study, the frequency of SHFM3 was 25%. The frequency of SHFM4 (*TP63* mutation) is an estimate from data of our own diagnostic laboratory (~15%) and a study of van Bokhoven et al.[48], where *TP63* mutations were detected in 4/35 (11%) of patients with isolated SHFM. The frequency of other SHFM types is until now not clear.

5. DISCUSSION

With MLPA and qPCR testing, we detected a duplication at the SHFM3 locus on chromosome 10q24 in 25% of the investigated patients. This is the first study where patients were selected only on basis of SHFM phenotype, without previous mapping to a specific locus. Syndromic and isolated cases were included, as well as familial and sporadic cases. This result confirms our hypothesis that the frequency of SHFM3 in SHFM patients is substantially high and, in fact, about two times higher than the (estimated) frequency of SHFM4 (*TP63* mutations), which is 10-15% according to the literature and our own experience.[48]

So far, there are two studies that have focused on the frequency of genomic rearrangements at the 10q24 (SHFM3) locus. Everman et al.[34] screened 44 cases of syndromic and non-syndromic forms of SHFM for duplications at the 10q24 locus, using pulse-field gel electrophoresis (PFGE) and real-time quantitative PCR (qPCR). Among these 44 cases eight chromosomal rearrangements were observed (18%), consisting of four sporadic cases, three families with autosomal dominant SHFM known to be linked to the SHFM3 locus and one non-linked family. De Mollerat et al.[28] detected tandem duplications at the SHFM3 locus in 7 families who were previously linked to this locus. Combining these studies, this brings the total of SHFM3-associated cases with chromosomal rearrangements to 15 of 51 (29%). This includes all of the cases that were previously linked to the SHFM3 locus (9 of 9 cases), and 6 of 42 additional cases (14%). Only 2 of the cases had syndromic forms of SHFM. If the previously linked cases are excluded, 6 of 42 cases (14%) show rearrangements at 10q24. The majority of these cases are non-syndromic forms of SHFM; thus SHFM3 seems to represent a substantial part of this phenotype. Relative frequency of SHFM3 from these studies was compared with *TP63* mutations (frequency was determined using estimates from the literature), where the frequency of SHFM3 rearrangements appears to exceed the frequency of *TP63* mutations (Figure 22).

5. Discussion

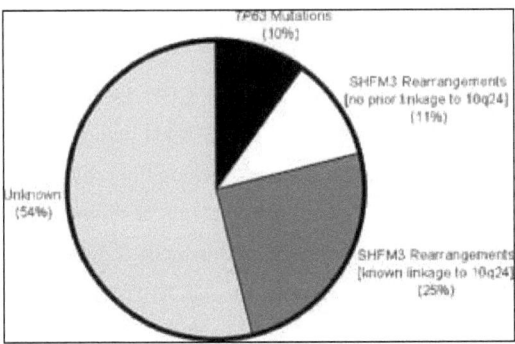

Figure 22. Relative frequencies of *TP63* mutations and SHFM3 rearrangements among non-syndromic SHFM cases. Adapted from: Everman et al. Frequency of genomic rearrangements involving the SHFM3 locus at chromosome 10q24 in syndromic and non-syndromic cases of split hand/foot malformation. Am J Med Genet 2006; 140A: 1375-1383.

Kano et al.[35] screened 28 Japanese families with non-syndromic cases of SHFM for duplications at the 10q24 locus using Southern blot and sequence analysis of the *DACTYLIN* gene. Of these 28 families, only 2 showed genomic rearrangements (7%). No mutations in the *DACTYLIN* gene were found. These families were not previously linked to the SHFM3 locus.

At first sight our results seem to be in concordance with the recent study of Everman et al.[34] who found SHFM3 rearrangements in 29% of all cases, including 9 of 9 cases (100%) with previous linkage to SHFM3. It is however difficult to compare these findings to our study, because 9 SHFM3-linked families were included. Kano et al.[35] described genomic rearrangements at the SHFM3 locus in only 2 families (7%). This result does not correlate with our findings and is unexpectedly low, especially considering the fact that only non-syndromic cases were included in contrast to our study, where syndromic cases were also included.

In the remaining 60% of the patients on which we performed MLPA, no copy number changes were detected at the known SHFM loci. However, recently evidence for two new SHFM loci was reported in the literature, on chromosome 8q [39] and a mutation in

                                                                    5. Discussion

WNT10b,[40] which we did not include in our study. It was unfortunately not possible to estimate the frequency of all SHFM types, because our cohort of 28 patients was too small to detect patients with rare types of SHFM, i.e. SHFM2. Therefore, I propose that the frequency of SHFM1, SHFM2 and SHFM5 must be lower than 3%.

In patient 1202, on whom I performed an array CGH to detect any pathological copy number changes, a deletion proximal to the SHFM1 locus was detected (illustrated schematically in Fig. 23). The deletion does not include the previously described candidate genes *DLX5*, *DLX6* or *DSS1*. Until now, only deletions or translocations further distal than the deletion in patient 1202 had been found at this locus.[17] There is strong evidence that, in mice, *Dlx5/6* and *Dss1* are responsible for the SHFM phenotype.[17,36] However, no mutations were detected in any of these genes, which has led to several theories about the pathomechanism of SHFM1. The array CGH findings in patient 1202 support the assumption that distant *cis*-acting regulatory elements are involved. When these elements are disrupted by the deletion which lies proximal to *DLX5*, *DLX6* and *DSS1*, this could result in a loss of function of these genes.

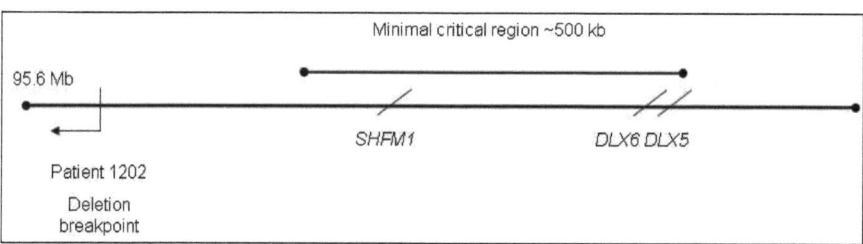

Figure 23. Location of SHFM1 critical region on chromosome 7q21. The minimal critical region for SHFM1, as designated by Crackower et al.,[17] comprises ~500kb and includes the candidate genes *SHFM1 (DSS1)*, *DLX5* and *DLX6*. The deletion detected in patient 1202 with array CGH lies proximal to this critical region (position: chr7:95139880-96582518 bp).

Of the seven SHFM3 patients I identified, none had associated syndromes, except for one case (patient 4774) with mild ectodermal involvement (oligodontia). This is consistent with a recent study, where ectodermal findings were most common in SHFM4 and SHFM3 patients showed only occasional nail or dental involvement.[2] However, it is possible that ectodermal involvement in this study's patients might be underestimated, because extensive clinical information was not available for all patients. Also, preaxial involvement

5. Discussion

of the upper extremities seems to be a significant discriminator for the 10q24 locus.[1] Most commonly seen are proximally placed thumbs, triphalangeal thumbs, preaxial polydactyly and/or absence of the first ray. One of the SHFM3 patients described here, patient 2269, indeed has preaxial unilateral polydactyly. But a total of seven patients in this study are too few to determine whether the preaxial involvement is a significant locus discriminator. Central clefting of the feet is also described as a significant discriminator for the SHFM3 locus.[1] With 4 of the SHFM3 patients from this study cohort showing clefting of the feet, 2 patients having monodactyly and one patient showing no foot involvement, this seems to be in concordance with my findings.

In this study I detected duplications of different sizes, ranging from ~435 kb to ~490 kb at the SHFM3 locus. Breakpoint estimation was performed with qPCR using different primers. The size of the duplication does not seem to correlate with the patient's phenotype, i.e. patient 536 with monodactyly of all four extremities carries a smaller duplication of about 440 kb, but in another patient with the same phenotype the duplication comprises about 490 kb. I therefore conclude that, in the group of SHFM 3 patients identified in this study, there is no correlation between the size of the duplication and the clinical phenotype. However, more research is needed with a larger cohort of SHFM3 patients to obtain a more significant result.

Since *TP63* mutation analysis is now a fairly common diagnostic test in most genetic laboratories, this is usually the first test requested when a SHFM patient presents to the clinical geneticist. Indeed, in patients with the EEC syndrome phenotype, a *TP63* mutation is identified in almost all patients.[48] However, as I have shown, in a cohort of 28 patients with SHFM in one or more limbs, 25% carry a duplication at the SHFM3 locus. Therefore I recommend that in clinical practice patients with isolated SHFM should primarily be tested for copy number changes at the SHFM3 locus by either MLPA or quantitative PCR.

MLPA and qPCR have proven to be effective methods to detect such aberrations. A great advantage of MLPA is that it is relatively fast and multiple loci can be tested in one experiment. However, one will only be able to find aberrations in the regions for which the probes are designed. For example, in patient 1202 a duplication was identified in a region proximal to the SHFM1 locus by array CGH. The MLPA probe for the SHFM1 locus, which lies in the SHFM1 region, would not have detected this aberration. Also, new loci such as that on chromosome 8q[36] cannot be included in the MLPA probe set because of the large

5. Discussion

interval (21 Mb). One should also take into account that designing and optimizing the probes for MLPA is a time-consuming process and experiments have to be repeated several times to get conclusive results. qPCR is a good method to detect copy number variations, and more detailed information can be obtained about the aberration breakpoints. The method is sensitive, fast, and reliable. A disadvantage is that, as in MLPA, primer design takes time and effort. Array CGH is an effective method to screen for copy number variations covering the whole genome. It is also possible to detect new aberrations, such as in the case of patient 1202. However array CGH is more expensive and therefore less suitable for routine testing.

FUTURE PROSPECTS

To date, five genetic loci are known to cause split hand/split foot malformation. Recently evidence for two new loci has been proposed. Looking at the fact that in 28 patients, I detected genomic aberrations at the known SHFM loci in only 25%, it is obvious that more genetic loci must exist which are responsible for this phenotype. Not only can these loci be detected by linkage analysis, which requires a large kindred, but also by using genome-wide scanning methods like array CGH to detect new copy number variations in sporadic SHFM patients. Since little is still known about the pathomechanism that leads to SHFM, both approaches will be a valuable tool for identifying further loci and candidate genes associated with this disorder.

By discovering more about the role of the candidate genes located at the different SHFM loci, we will be able to understand more about the developmental pathways that are deregulated in SHFM patients. Animal models can be a helpful tool in determining the function of these candidate genes and their role in limb morphogenesis.

ZUSAMMENFASSUNG

Split hand/split foot malformation (SHFM) ist eine hauptsächlich autosomal dominant vererbte Skeletterkrankung, wobei sowohl Hände als auch Füße betroffen sein können. Sie wird gekennzeichnet durch eine tiefe mediane Spalte mit Aplasie der zentralen Strahlen. Diese Erkrankung tritt sowohl sporadisch als auch familiär auf. Ein typisches Merkmal ist die inter- und intrafamiliäre Variabilität, wobei der Phänotyp variiert von isolierter häutiger Syndaktylie bis zur schwersten Form der SHFM, der Monodaktylie. Häufig treten auch zusätzliche phänotypische Merkmale auf wie triphalangealer Daumen, Polydaktylie oder Klinodaktylie. In 40% der Fälle liegen weitere nicht-skelettale Fehlbildungen vor. Bisher sind fünf genetische Loki identifiziert worden, die zu der Entstehung der SHFM beitragen. Vor kurzem sind darüber hinaus zwei mögliche neue Loki entdeckt worden. SHFM1 (MIM 183600) ist assoziiert mit chromosomalen Aberrationen in der Region 7q21-q22. Kandidatengene in dieser Region sind *DLX5*, *DLX6*, und *DSS1*. Eine X-chromosomal rezessive Vererbung der SHFM auf Chrosomosom Xq26 wurde in nur einer Familie beobachtet (SHFM2, MIM 313350). Ein dritter Lokus (SHFM3, MIM 600095) befindet sich auf Chromosom 10q24. Hier wurde eine Duplikation nachgewiesen, die variabel ist zwischen einer Größe von ~325 kb bis ~500 kb. In den Kandidatengenen in dieser Region (*BRTC*, *POLL* und *LBX1*) konnten bisher keine Mutationen festgestellt werden. Bei SHFM4 (MIM 605289) jedoch sind Mutationen in dem Gen *TP63* (Chromosom 3q27) assoziiert mit dem Phänotyp. *TP63* Mutationen sind nicht nur bei isolierten Formen der SHFM, sondern vor allem bei Syndrom-assoziierter SHFM, sowie dem Ectrodactyly-ectodermal Dysplasia-Clefting (EEC) Syndrom beschrieben worden. Der Lokus für SHFM5 (MIM 606708) liegt auf Chromosom 2q31 und umfasst zwei Kandidatengene *DLX1* and *DLX2*.

TP63 Mutationsanalyse wird zurzeit als Routine-Test bei der SHFM Diagnostik durchgeführt. Jedoch gibt es Hinweise, dass die Frequenz von SHFM3 wesentlich höher ist, als die von SHFM4. Es gibt bisher noch keine eindeutigen Studien, in denen die Frequenz der verschiedenen Typen SHFM, ausschließlich basierend auf dem Phänotyp, bestimmt wurde. In dieser Arbeit wurden 28 Patienten mit SHFM in einer oder mehreren Extremitäten untersucht, wobei sowohl sporadische als auch familiäre Fälle eingeschlossen wurden. Patienten, bei denen eine *TP63*-Mutation gefunden wurde, wurden von den weiteren Untersuchungen ausgeschlossen. Mit *Multiplex Ligation-dependent Probe Amplification* (MLPA) wurden Kopienzahl-Veränderungen in allen fünf

Zusammenfassung

bekannten SHFM Loki untersucht. Es wurden sieben Patienten mit einer Duplikation auf dem SHFM3 Lokus identifiziert. Diese Aberrationen konnten mit quantitativer Polymerase-Kettenreaktion (qPCR) bestätigt werden. An den anderen vier Loki wurden keine Kopienzahl-Veränderungen nachgewiesen. D.h. die Frequenz von SHFM3 in der hier untersuchten SHFM-Kohorte ist 25% (7/28). Dies ist fast doppelt so hoch wie die Frequenz von SHFM4 (10-15%). Aufgrund dieser Daten empfiehlt sich für die klinische Praxis, Patienten mit isolierter SHFM zunächst auf Kopienzahl-Veränderungen am SHFM3 Lokus (Chromosom 10q24) zu untersuchen. MLPA und qPCR sind effektive Methoden, um diese Aberrationen nachzuweisen. Für syndromal-assoziierte SHFM Patienten wäre es sinnvoll, als erste diagnostische Unterschung eine *TP63*-Mutationsanalyse durchzuführen.

REFERENCES

1. Elliott AM, Reed MH, Roscioli T, Evans JA. Discrepancies in upper and lower limb patterning in split hand foot malformation. Clin Genet 2005; 68: 408-423.

2. Elliott AM, Evans JA. Genotype-Phenotype Correlations in Mapped Split Hand Foot Malformation (SHFM) Patients. Am J Med Gen 2006; Part A 140A: 1419-1427.

3. Sifakis S, Basel D, Ianakiev P, Kilpatrick MW, Tsipouras P. Distal limb malformations; underlying mechanisms and clinical associations. Clin Genet 2001; 60: 165-172.

4. Basel D, Kilpatrick MW, Tsipouras P. The expanding Panorama of Split Hand Foot Malformation. Am J Med Gen 2006; Part A 140A: 1359-1365.

5. McKusick V. Mendelian Inheritance in Man: A Catalog of Human Genes and Genetic Disorders. 12th Ed Baltimore, USA: John Hopkins University Press, 1998.

6. Elliott AM, Evans JA, Chudley AE. Split hand/foot malformation (SHFM). Clin Genet 2005; 68: 501-505.

7. Elliott AM, Reed MH, Chudley AE, Chordirker BN, Evans JA. Clinical and epidemiological Findings in Patients with Central Ray Deficiency: Split Hand Foot Malformation. Am J Med Gen 2006; Part A 140A: 1428-1439.

8. Lange M. Grundsätzliches über die Beurteilung der Entstehung und Bewertung atypischer Hand- und Fussmissbildungen. Z Orthoped 1937; 66: 80.

9. Barsky AJ. Cleft hand: classification, incidence, and treatment. Review of the literature and report of nineteen cases. J Bone Joint Surg Am 1964; 46: 1707-1720.

10. Birch-Jensen A. Congenital deformities of the upper extremities. Copenhagen: Det. Danske Forlag, 1949.

11. Manske PR. Symbrachydactyly instead of atypical cleft hand. Plast Reconstr Surg 1993; 91: 196.

12. Gül D, Öktenli Ç. Evidence for autosomal recessive inheritance of split hand/split foot malformation: a report of nine cases. Clin Dysmorph 2002; 11(3): 183-186.

13. De Smet L, Fabry G. Characteristics of patients with symbrachydactyly. J Pediatr Orthop B 1998; 7(2): 158-161.

14. Schouten JP, McElgunn CJ, Waaijer R et al. Relative quantification of 40 nucleid acid sequences by multiplex ligation-dependent probe amplification. Nucleid Acids Research 2002; 30(12): e57.

15. Scherer SW, Poorkaj P, Massa H et al. Physical mapping of the split hand/split foot locus on chromosome 7 and implication in syndromic ectrodactyly. Hum Mol Gen 1994; 3(8): 1345-1354.

16. Scherer SW, Poorkaj P, Allen T et al. Fine mapping of the autosomal dominant split hand/split foot locus on chromosome 7, band q21.3-q22.1. Am J Hum Genet 1994; 55:12-20.

17. Crackower MA, Scherer SW, Rommens JM et al. Characterization of the split hand/split foot malformation locus SHFM1 at 7q21.3-q22.1 and analysis of a candidate gene for its expression during limb development. Hum Mol Gen 1996; 5(5): 571-579.

18. Ahmad M, Abbas H, Haque S et al. X-chromosomally inherited split-hand/split-foot anomaly in a Pakistani kindred. Hum Genet 1987, 75: 169-173.

19. Faiyaz ul Haque M, Uhlhaas S, Knapp M et al. Mapping of the gene for X-chromosomal split-hand/split-foot anomaly to Xq26-q26.1.Hum Genet 1993; 91: 17-19.

20. Friedli M, Nikolaev S, Lyle R et al. Characterization of mouse Dactylaplasia mutations: a model for human ectrodactyly SHFM3. Mamm Genome 2008; 19: 272-278.

21. Ianakiev P, Kilpatrick MW, Toudjarska I et al. Split hand/split foot malformation is caused by mutations in the p63 gene on 3q27. Am J Hum Gen 2000; 67(1): 59-66.

22. Del Campo M, Jones MC, Veraksa AN et al. Monodactylous limbs and abnormal genitalia are associated with hemizygosity for the human 2q31 region that includes the HOXD cluster. Am J Hum Genet 1999; 65(1): 104-110.

23. Merlo GR, Zerega B, Paleari L et al. Multiple functions of Dlx genes. Int J Dev Biol 2000; 44(6): 619-626.

24. Duijf PHG, van Bokhoven H, Brunner HG. Pathogenesis of split-hand/split-foot malformation. Hum Mol Genet 2003; 12: R51-60.

25. Schwabe GC, Mundlos S. Genetics of congenital hand anomalies. Handchir Mikrochir Plast Chir 2004; 36: 85-97.

26. Crackower MA, Motoyama J, Tsui L. Defect in the maintenance of the Apical Ectodermal Ridge in the Dactylaplasia mouse. Dev Biol 1998; 201: 78-89.

27. Sidow A, Bulotsky MS, Kerrebrock AW et al. A novel member of the F-box/WD40 gene family, encoding dactylin, is disrupted in the mouse dactylaplasia mutant. Nat Genet 1999; 23: 104-107.

28. De Mollerat XJ, Gurrieri F, Morgan CT et al. A genomic rearrangement resulting in a tandem duplication is associated with split hand-split foot malformation 3 (SHFM3) at 10q24. Hum Mol Genet 2003; 12(16): 1959-1971.

29. Boles RG, Pober BR, Gibson LH et al. Deletion of chromosome 2q24-q31 causes characteristic digital anomalies: case report and review. Am J Med Gen 1995; 55(2): 155-160.

30. Elliot AM, Reed MH, Evans JA. Triphalangeal thumb in association with split hand/foot: a phenotypic marker for SHFM3? Birth Defects Res A Clin Mol Teratol 2007; 79: 58-61.

31. Majewski F, Küster W, ter Haar B et al. Aplasia of tibia with split-hand/split-foot deformity: report of six families with 35 cases and considerations about variability and penetrance. Hum Gen 1985; 70: 136-147.

32. Babbs C, Heller R, Everman DB et al. A new locus for split hand/foot malformation with long bone deficiency (SHFLD) at 2q14.2 identified from a chromosome translocation. Hum Genet 2007; 122: 191-199.

33. Evans JA, Reed MH, Greenberg CR. Fibular aplasia with ectrodactyly. Am J Med Genet 2002; 113: 52-58.

34. Everman DB, Morgan CT, Lyle R et al. Frequency of genomic rearrangements involving the SHFM3 locus at chromosome 10q24 in syndromic and non-syndromic split hand/foot malformation. Am J Med Genet 2006; 140A: 1375-1383.

35. Kano H, Kurosawa K, Horii E et al. Genomic rearrangement at 10q24 in non-syndromic split-hand/split-foot malformation. Hum Genet 2005; 118: 477-483.

36. Robledo RF, Lakshmi R, Li X, Lufkin T. The Dlx5 and Dlx6 homeobox genes are essential for craniofacial, axial, and appendicular skeletal development. Genes Dev 2002; 16(9): 1089-1101.

37. Tackels-Horne D, Toburen A, Sangiorgi E et al. Split hand/foot malformation with hearing loss: first report of families linked to the SHFM1 locus in 7q21. Clin Genet 2001, 59: 28-36.

38. Bernardini L, Palka C, Ceccarini C et al. Complex rearrangement of chromosome 7q21.13-q22.1 confirms the ectrodactyly-deafness locus and suggests new candidate genes. Am J Med Genet 2008; 146A: 238-244.

39. Gurnett CA, Dobbs MB, Nordsieck EJ et al. Evidence for an additional locus for split hand/split foot malformation in chromosome region 8q21.11-q22.3. Am J Med Genet 2006; 140A: 1744-1748.

40. Ugur SA, Tolun A. Homozygous WNT10b mutation and complex inheritance in Split-Hand/Foot Malformation. Hum Mol Genet 2008; 17(17): 2644-2653.

41. Nunes ME, Schutt G, Kapur RP et al. A second autosomal split hand/split foot locus maps to chromosome 10q24-25. Hum Mol Genet 1995; 4(11): 2165-2170.

42. Gurrieri F, Prinos P, Tackels D et al. A split hand-split foot (SHFM3) gene is located at 10q24→25. Am J Med Genet 1996; 62(4): 427-436.

43. Ianakiev P, Kilpatrick MW, Dealy C et al. A novel human gene encoding an F-box/WD40 containing protein maps in the SHFM3 critical region on 10q24. Biochem. Biophys. Res. Commun. 1991; 261: 64-70.

44. Goodman FR, Majewski F, Collins AL, Scambler PJ. A 117-kb microdeletion removing HOXD9-HOXD13 and EVX2 causes synpolydactyly. Am J Hum Genet 2002; 70: 547-555.

45. Qiu M, Bulfone A, Ghattas I et al. Role of the Dlx Homeobox Genes in Proximodistal Patterning of the Branchial Arches: Mutations of Dlx-1, Dlx-2, and Dlx-1 and -2 Alter Morphogenesis of Proximal Skeletal and Soft Tissue Structures Derived from the First and Second Arches. Dev Biol 1997; 185 (2): 165-184.

46. Celli J, Duijf P, Hamel BCJ et al. Heterozygous germline mutations in the p53 homolog p63 are the cause of EEC syndrome. Cell 1999; 99: 143-153.

47. Mills AA, Zheng B, Wang X et al. P63 is a p53 homologue required for limb and epidermal morphogenesis. Nature 1999; 398: 708-713.

48. Van Bokhoven H, Hamel BCJ, Bamshad M et al. P63 gene mutations in EEC syndrome, limb-mammary syndrome, and isolated split hand-split foot malformation suggest a genotype-phenotype correlation. Am J Hum Genet 2001; 69: 481-492.

49. Lo Iacono N, Mantero S, Chiarelli A et al. Regulation of Dlx5 and Dlx6 gene expression by p63 is involved in EEC and SHFM congenital limb defects. Development 2008; 135: 1377-1388.

50. Lyle R, Radhakrishna U, Blouin J et al. Split-Hand/Split-Foot Malformation 3 (SHFM3) at 10q24, development of rapid diagnostic methods and gene expression from the region. Am J Med Genet 2006; 140A: 1384-1395.

51. Basel D, DePaepe A, Kilpatrick MW et al. Split hand foot malformation is associated with a reduced level of Dactylin gene expression. Clin Genet 2003; 64: 350-354.

52. Oostlander AE, Meijer GA, Ylstra B. Microarray-based comparative genome hybridization and its applications in human genetics. Clin Genet 2004; 66(6): 488-495.

53. Lockwood WW, Chari R, Chi B et al. Recent advantages in array comparative genomic hybridization technologies and their applications in human genetics. Eur J Hum Genet 2006; 14(2): 139-148.

54. Weimer J, Kiechle M, Wiedemann U et al. Delineation of a complex karyotipic rearrangement by microdissection and CGH in a family affected with split foot. J Med Genet 2000; 37: 442-445.

Die VDM Verlagsservicegesellschaft sucht für wissenschaftliche Verlage abgeschlossene und herausragende

Dissertationen, Habilitationen, Diplomarbeiten, Master Theses, Magisterarbeiten usw.

für die kostenlose Publikation als Fachbuch.

Sie verfügen über eine Arbeit, die hohen inhaltlichen und formalen Ansprüchen genügt, und haben Interesse an einer honorarvergüteten Publikation?

Dann senden Sie bitte erste Informationen über sich und Ihre Arbeit per Email an *info@vdm-vsg.de*.

Sie erhalten kurzfristig unser Feedback!

VDM Verlagsservicegesellschaft mbH
Dudweiler Landstr. 99
D - 66123 Saarbrücken
www.vdm-vsg.de

Telefon +49 681 3720 174
Fax +49 681 3720 1749

Die VDM Verlagsservicegesellschaft mbH vertritt

Printed by Books on Demand GmbH, Norderstedt / Germany